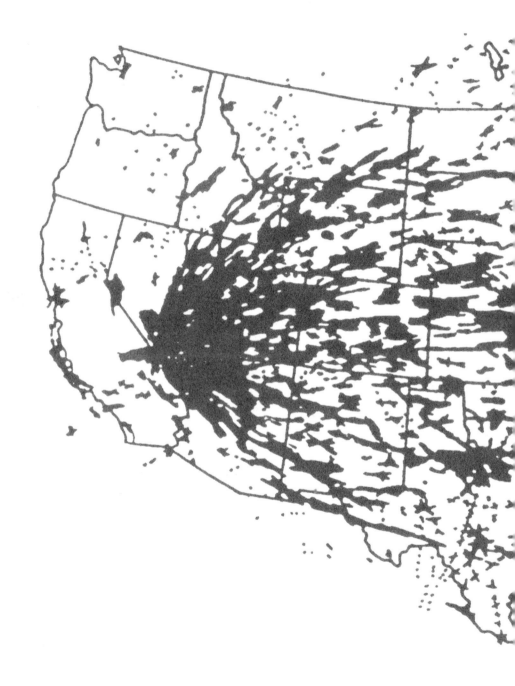

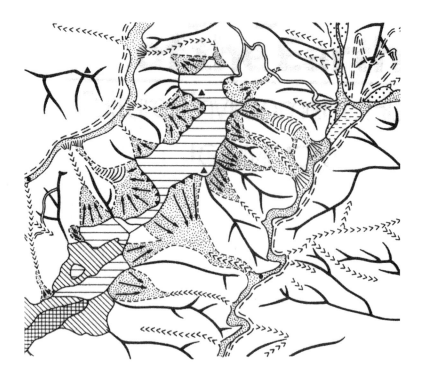

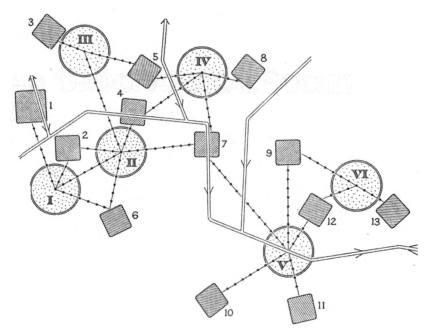

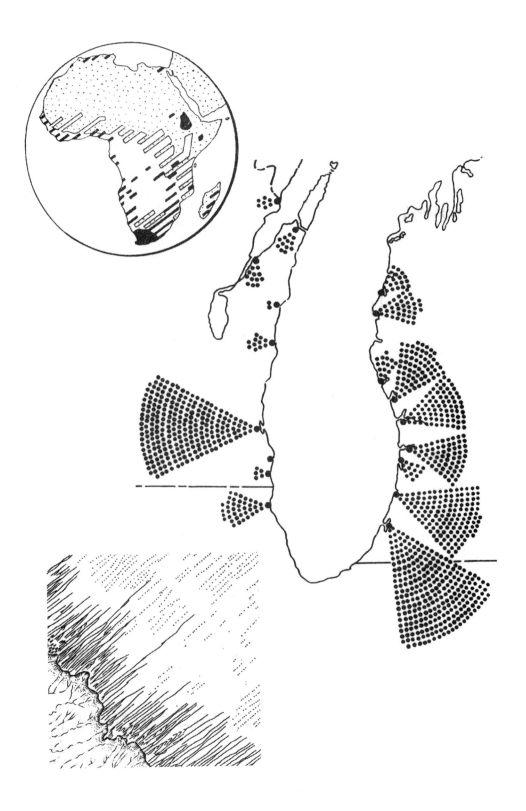

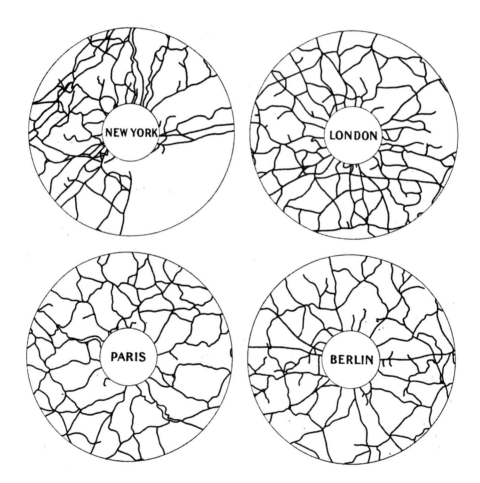

# making
## maps

### a visual guide to
### map design for gis

## john krygier
## denis wood

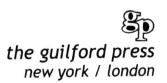

the guilford press
new york / london

© 2005 The Guilford Press
A Division of Guilford Publications, Inc.
72 Spring Street, New York, NY 10012
www.guilford.com

Printed in the United States of America

This book is printed on acid-free paper.

Last digit is print number:   9   8   7   6   5   4   3   2   1

**Library of Congress Cataloging-in-Publication Data**

Krygier, John.
  Making maps : a visual guide to map design for GIS / John Krygier,
  Denis Wood.
     p.   cm.
   Includes index.
   ISBN 1-59385-200-2 (pbk.)
  1. Cartography.  2. Geographic information systems.   I. Wood, Denis.
II. Title.
GA105.3K79   2005
526—dc22

                                          2005018239

*This book is dedicated to, and in memory of, David Woodward, map designer and scholar*

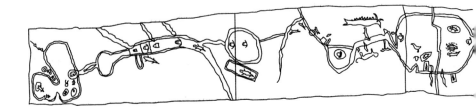

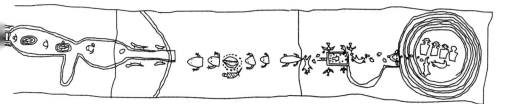

*What is it?*

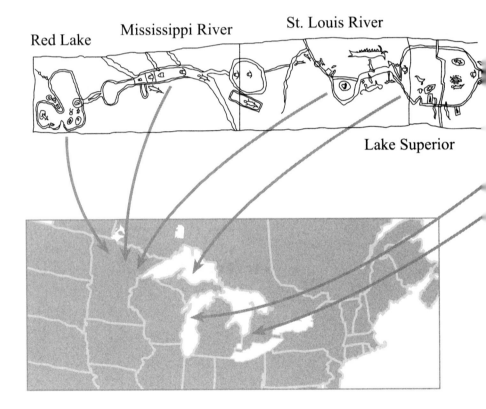

Red Lake

Mississippi River

St. Louis River

Lake Superior

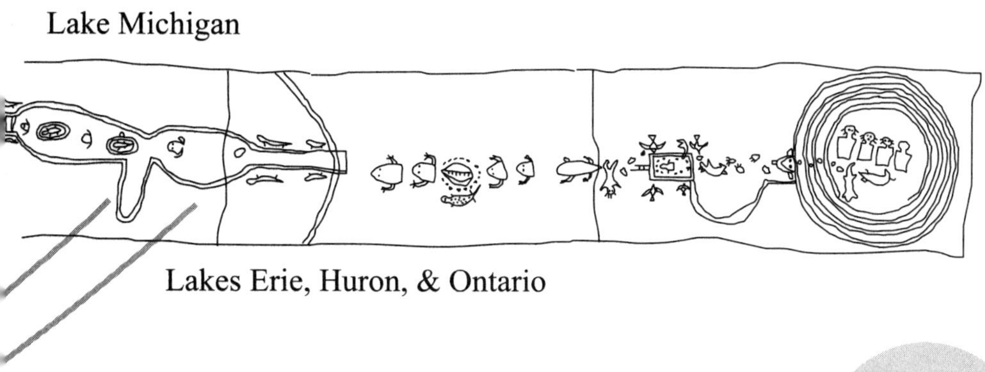

Lake Michigan

Lakes Erie, Huron, & Ontario

# It's a Map

## Ojibwe (Native American) ca. 1820

Maps are a powerful way of thinking about the earth.

This Native map, drawn on birch bark (which accounts for its shape), shows the migration legend of the Ojibwe, from the creation of their people (on the right) to their home in the upper Midwest (on the left). The left and central portions of the map show Lake Huron, Lake Superior, and Red Lake in Minnesota. The right side of the map relates the spiritual realities of the Ojibwe origins with important spiritual guides symbolized along the route. The map is a sophisticated synthesis of spiritual and physical geography, revealing the vital importance of making maps in the context of your life and belief systems.

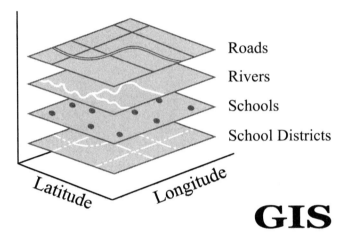

Roads

Rivers

Schools

School Districts

Latitude

Longitude

# GIS

# Technological Changes

## Geographic Information Systems (GIS)

Technological changes have made the mapping and analysis of geographic information a daily part of how we understand the world. Geographic Information Systems (GIS) consist of digital map layers that store a multitude of geographic facts by location: water features, property parcels, floodplains, roads, political boundaries, and so on. Digital map layers are linked to a database of "attribute" information. Each property parcel, for example, contains information such as the owner, address, zoning, and value of the property. Each stream and river contains information including the name of the feature, its length, and flow.

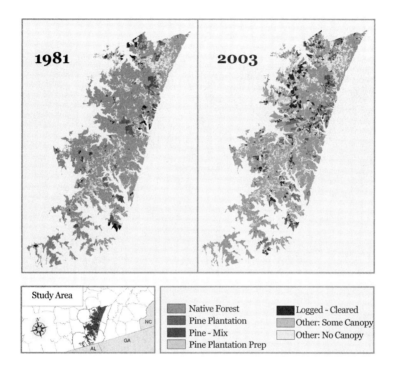

1981

2003

Study Area

NC

GA

AL

Native Forest
Pine Plantation
Pine - Mix
Pine Plantation Prep

Logged - Cleared
Other: Some Canopy
Other: No Canopy

## Forest change in Tennessee

GIS is an excellent way to inventory geographic facts, but its power lies in its analytical capabilities. Jon Evans of Sewanee University used GIS to analyze forest change in Tennessee. Combining satellite data and GIS map layers, Evans' GIS analysis revealed a 14% loss in native forests and a 170% increase in pine plantations between 1981 and 2000. The *quality* of forests is a politically sensitive issue. The lumber and paper industry and certain U.S. government policy-makers claim U.S. forests are increasing in area. Biologists argue that diverse native forests are being replaced by dangerous monoculture tree plantations: not all trees make a forest. Evans' GIS analysis and the resulting maps provide quantitative and visual evidence for the growing debate over forest quality and policy in the U.S.

# Maps Kill!

## Military targeting maps

Maps have long been used to fight wars, win battles, and kill people.

Targeting maps help to fine-tune bombing during military campaigns, to destroy certain select targets and people and not others. Royal Air Force pilots are briefed with a terrain model prior to a bombing mission during World War II (top). Geographic Information Systems – and maps of bombing targets – are at the core of strategizing and planning bombing missions for the war in Iraq (bottom).

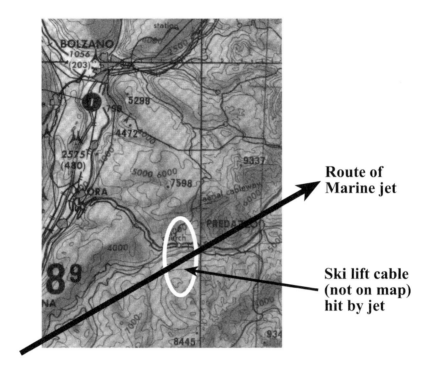

**Route of Marine jet**

**Ski lift cable (not on map) hit by jet**

# 20 killed near Cavalese, northern Italy

Despite a long tradition of map making, and expensive and extensive new technologies for making maps, maps still can fail and so kill.

A U.S. Marine Corps jet struck and severed a ski lift cable spanning an Alpine valley in northern Italy in 1998, sending the ski lift gondola crashing to the earth, killing 20 people.

The jet crew did have a map of the area, but not one that showed the ski lift cable. The cable was shown on Italian maps, but the Pentagon prohibits the use of maps made by foreign nations.

Map users and makers should always be critical of maps: maps have the power and potential for failing that every human-created object has.

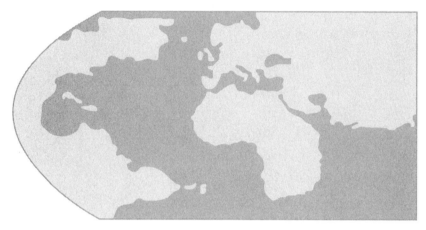

*Juan de la Cosa's map of the known world,
from the early 16th century.*

# Discoveries with Maps

## Maps generate new understanding

Maps don't simply locate things: they can lead us to insights, discoveries, and new ways of understanding.  We see geographic patterns, and those patterns may lead us to think about things in a new way.

When maps of the then known world were first made in the 16th century, people immediately noticed provocative patterns: the earth's continents seemed to fit together like pieces of a puzzle. Many explanations for these patterns were put forth for debate.

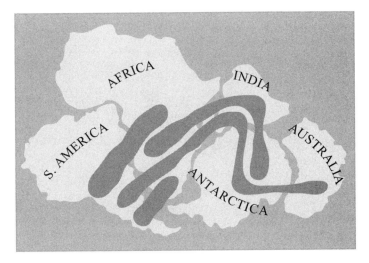

*Geologic and fossil commonalities in dark grey,
spanning different "continents" in Pangea.*

## Plate tectonics

Geologic evidence gathered by Alfred Wegener and others suggested
that a single large continent, Pangea, had indeed preceded the current
configuration of the continents. Evidence included geologic and fossil
characteristics common to continents now widely separated. But how
did the continents move?

Geologic research in the 1950s and 1960s led to the understanding that
molten rock from deep inside the earth pushes up along major cracks
in the earth's crust, building new crust, and pushing and moving the
huge plates, upon which the continents rest. Plate tectonics is now
accepted as the explanation for the continental patterns on maps that
intrigued humans for so long.

Map making is powerful, as it can lead us to discoveries of new things
about the natural and human world.

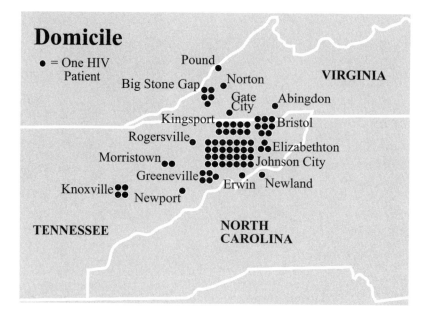

**Domicile**

● = One HIV Patient

Pound
Norton
Big Stone Gap
Gate City
Abingdon
**VIRGINIA**
Kingsport
Bristol
Rogersville
Elizabethton
Morristown
Johnson City
Greeneville
Erwin   Newland
Knoxville
Newport

**TENNESSEE**

**NORTH CAROLINA**

# Thinking Visually with Maps

## HIV patients in rural Tennessee

Maps are a powerful way to think things through. Abraham Verghese used maps to help think about his HIV-infected patients in the mid-1980s. Dr. Verghese practiced medicine in rural Tennessee. He and his colleagues were stunned when HIV-infected patients began to dominate their practices. What was this urban problem doing in rural Tennessee? "There was a pattern in my HIV practice. I kept feeling if I could concentrate hard enough, step back and look carefully, I could draw a kind of blueprint that explained what was happening here..." Dr. Verghese borrowed a map of the U.S. from his son. With the map spread on his living room floor, he marked where his HIV patients lived. He labeled the map **Domicile**, but he could as well have called it "Birthplace," for most of his patients were men who had come home to die.

# Acquisition

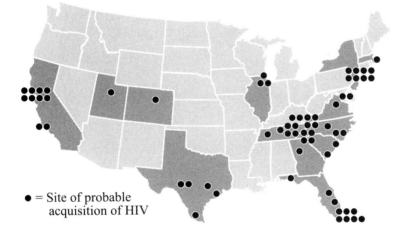

● = Site of probable
   acquisition of HIV

Dr. Verghese next mapped where his HIV patients lived between 1979 and 1985. The places on the **Acquisition** map "seemed to circle the periphery of the United States" and were mostly large cities. "As I neared the end, I could see a distinct pattern of dots emerging on this larger map of the USA. All evening I had been on the threshold of seeing. Now I fully understood." Dr. Verghese learned of a circuitous voyage, a migration from home and a return, ending in death. It was "the story of how a generation of young men, raised to self-hatred, had risen above the definitions that their society and upbringings had used to define them. It was the story of hard and sometimes lonely journeys they took far from home into a world more complicated than they imagined and far more dangerous than anyone could have known." Patients that appear on both maps are those "who had the virus delivered to their doorstep... hemophiliacs or blood transfusion recipients who got tainted ... blood products." The maps Verghese made on his living room floor might not be much to *look* at, but the *thinking* they inspired was rich.

# Maps Shape How We See

## The earth is really big and complex

Maps are small and show only a few of the multitude of human and natural features. When making maps, we strip away selected details and flatten the earth's surface, showing what we could not otherwise see.

## Less detail

Map makers remove detail to show what they choose to show.

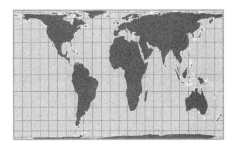

## Entire earth, all at once

Map makers flatten the earth's entire surface. This map stretches continental shapes, revealing distortions that occur when we flatten the earth's surface.

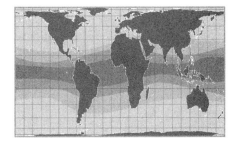

## Seeing the invisible

On maps we can record what is visible to us – coastlines – and what is not visible to us – temperatures.

## The impossible is natural

Global temperature is something we could never "see" without maps. Nevertheless, such an impossible "view" of the world is now quite natural to us: our maps shape how we "see" the world.

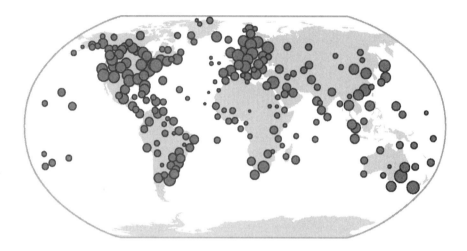

*The geography of the internet:*
*larger circles = more internet hosts.*

# The internet

## The internet lets anyone make maps

Changes in technology have always affected map making. The invention of printing made it easy to duplicate and distribute maps, providing more people with access to maps and the knowledge they store. The internet is having a profound impact on map access. Free internet mapping sites provide both basic and sophisticated map-making capabilities. With internet access, and the ability to point and click, anyone can make maps.

## Where is ...

Where is 231 Crestview Road, in Columbus, Ohio?

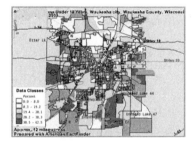

## What is the route ...

How do I get from 231 Crestview Road to Delaware, Ohio?

## How many ...

How many people live in Delaware, Ohio, and where are they?

## Making maps is ubiquitous!

Map making is part of our day-to-day lives.

*Areas crossed by two or more radioactive clouds during the era of nuclear testing in the American Southwest, 1951-62.*

# What Is a Map?

## A graphic statement that locates facts

**Graphic:** A visual display of marks which stand for something else. An airplane shape on a map implies an airport.

**Statement:** To put forth information, a formal embodiment of facts or assertions.

**Locating facts:** Tangible and intangible phenomena located in geographic space: what we can see (roads, rivers) and what we cannot see (temperature, radiation), varying in amount (population) and kind (vegetation types).

# Making Maps

## How do you make a map?

**Why are you making your map?** What are you intending to do with it? Who is your audience? Are they experts or novices? Young or old? What should the map assert? What do you want your map to communicate to those who see it? You must decide *why* you are making the map before you make it. *Chapter 2.*

**Mappable data:** Given your purpose, what data do you need? Roads? Rivers? Coastlines? Population? Toxin levels in wells? Number of deer per county? Some data are free and easy to find. Other data exist, but at a cost. The facts you need may not exist at all, and you may have to collect them yourself or pay someone to do so. Locating and processing mappable data can be among the most expensive and time-consuming parts of making maps. *Chapter 3.*

**Map-making tools:** Which tools will you use to make your map? Internet map-making sites are free but limited in data and map-making capabilities. Geographical Information System (GIS) software is more sophisticated and flexible but expensive to purchase and more difficult to use. Select appropriate map-making tools based on the kind of map you want to make: you cannot make every map using internet sites, but you don't need GIS for every map either. *Chapter 4.*

**Map design:** Given the reason you are making the map, mappable data, and map-making tools, how do you make a map that fulfills your goals - a map that works? Key issues include the geographic framework, map layout, intellectual and visual hierarchies, map generalization and classification, map symbolization, and the use of type and color. *Chapters 5-12.*

"Who died and made you the map police?"

Jill. *Home Improvement.* (TV, 1991)

"Is that the same map?" Jincey asked. She pointed to the large map of the world that hung, rolled up for the summer, above the blackboard behind Miss Dove. "Is China still orange?"

"It is a new map," Miss Dove said. "China is purple."

"I liked the old map," Jincey said. "I liked the old world."

"Cartography is a fluid art," said Miss Dove.

Frances Gray Patton. *Good Morning, Miss Dove.* (1954)

# more information...

Three excellent books situate mapping in its broader human context: Daniel Dorling and David Fairbairn, *Mapping* (Longman, 1997); Mark Monmonier, *How to Lie with Maps* (2nd ed., University of Chicago Press, 1996); and Denis Wood, *The Power of Maps* (Guilford Press, 1992). Check out the journal *Cartographic Perspectives;* published by the North American Cartographic Information Society.

This book draws from numerous texts, which can be consulted for more information: Borden Dent, *Cartography: Thematic Map Design* (5th ed., McGraw-Hill, 1998); Alan MacEachren, *Some Truth with Maps* (Association of American Geographers, 1994); and Phillip and Juliana Muehrcke, *Map Use* (4th ed., JP Publications, 1998). Also consulted were M.J. Kraak and F.J. Ormeling, *Cartography: Visualization of Spatial Data* (Longman, 1996); Arthur Robinson et al., *Elements of Cartography* (6th ed., Wiley, 1995); and Terry Slocum et al., *Thematic Cartography and Visualization* (2nd ed., Prentice Hall, 2003). These folks are the map police.

An older generation of mapping textbooks are worth looking at: F.J. Monkhouse and H.R. Wilkinson, *Maps and Diagrams* (Methuen, 1952); J.S. Keates, *Cartographic Design and Production* (Wiley, 1973); and two wonderful textbooks by Erwin Raisz: *General Cartography* (McGraw-Hill, 1938) and *Principles of Cartography* (McGraw-Hill, 1962).

The most comprehensive overview of academic research on mapping is Alan MacEachren, *How Maps Work* (Guilford Press, 1995).

**Sources:** The Ojibwe map (pp. xii, 1-3) redrawn from Selwyn Dewdney, *The Sacred Scrolls of the Southern Ojibway* (University of Toronto Press, 1975). The forest change map (p. 5) is courtesy of Jon Evans. The RAF terrain model image (p. 6) is from the *Geographical Review,* Oct. 1946. The Iraq military targeting map (p. 6) was found floating on the internet. The Cavalese map (p. 7) is courtesy of the Harvard Map Library, Harvard University. AIDS data (pp. 10-11) from Abraham Verghese, *My Own Country: A Doctor's Story* (Vintage, 1994) and "Urbs in Rure: Human Immunodeficiency Virus Infection in Rural Tennessee" (*Journal of Infectious Diseases,* 160:6). The internet hosts map (p. 14) is based on one from Matrix Net Systems. Internet road maps from Mapsonus.com, and U.S. Census map from the U.S. Census American Factfinder. The radioactive cloud map is from Richard L. Miller, *Under the Cloud: The Decades of Nuclear Testing* (Two-Sixty Press, 1999).

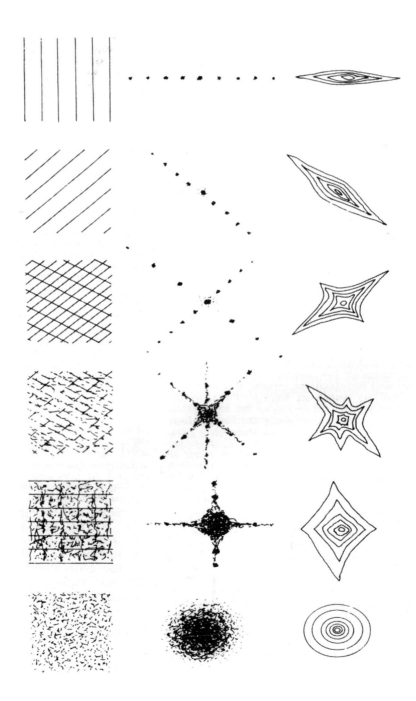

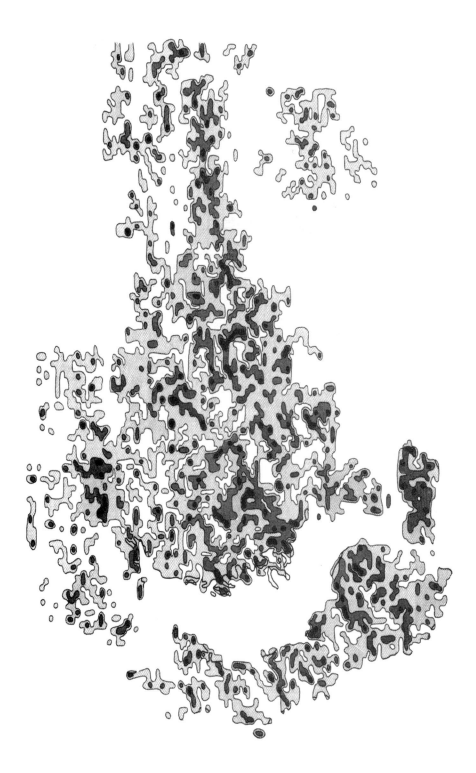

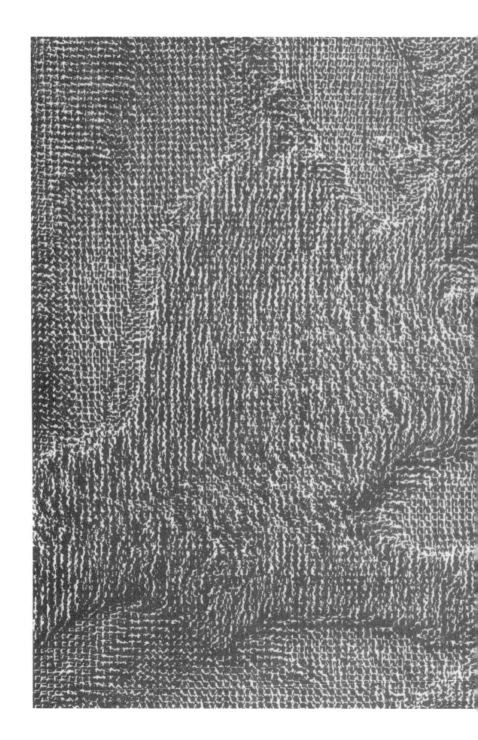

*Whose purpose
does it serve?*

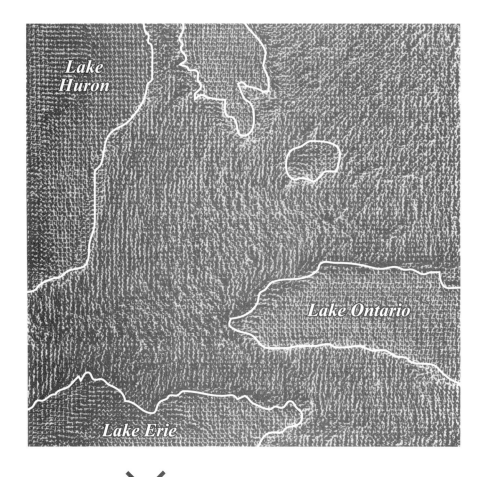

The many small stick-figures vary, depending on the amount of energy reflected at a particular location.

# Why Are You Making Your Map?

***Maps look like what they are supposed to do.*** Each mark on this map is a complex symbol representing five different types of energy reflected from the earth's surface. The map shows an area in Ontario, including parts of Lakes Huron, Erie, and Ontario. Most people won't be able to make much sense out of this map, but to expert earth scientists the complex symbols and specialist subject matter are necessary.

# Why Are You Making Your Map?

What are you trying to say with your map?  Who are you saying it to?
What do they know?  How will they use it?  Are they going to see it on
a computer, paper, poster, or projected on a screen during a presentation?
Careful consideration of these issues will guide the making of your map
and will produce a map that more effectively accomplishes what you
want it to do.

**Why are you making your map?**
Prior to making a map, clarify your
intent: intent shapes design.

**Who is your map for?**
Knowing the intended audience for your
map will help you design it.

**What is the final medium?**
The final form of your map - paper,
projected, etc. - will affect its design.

**Evaluating your map**
Evaluation plays a useful
role in making maps.

# Why are you making your map?

Prior to making a map, clarify your intent. Simply writing out the purpose of the map prior to making it will clarify goals; help determine relevant data, map design, and symbolization choices; and will lead to a better map.

**What the map is for:** A map showing a proposed Black Heritage Trail in Eli County, VA. The map is the visual centerpiece of a proposal for grants to develop the trail and its associated sites, and must visually tantalize granting agencies.

Poor:

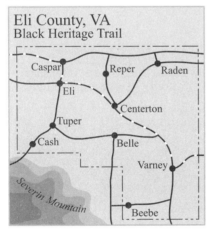

Good:

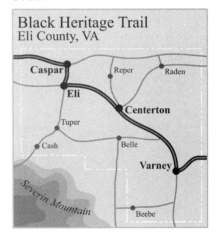

✓ title suggests county rather than trail as primary subject of the map.

✓ hard to figure out where the trail is.

✓ cities and roads along trail not visually different from other cities and roads.

✓ little visual depth to the map: trail is not visually prominent.

✓ title suggests trail as primary subject of the map.

✓ easy to see the trail.

✓ cities and roads along trail are visually prominent.

✓ meaningful visual depth to the map: trail is visually prominent.

**Different goals call for different maps!** Frequently the quality of a map is a matter of perspective, not design. This is because a map is a statement locating facts, and people tend to select the facts that make their case. That's what the map is for: to make their case.

Consider the examples below. A proposed connector road (dashed black) cuts through a city. Different groups create equally good maps to articulate their different perspectives on the proposed route. Though the maps may seem polemical, isolating the facts each presents is useful in focusing debate.

**Goal:** The County Chamber of Commerce shows the shortest and least costly route for the connector. They focus on property values:

**Goal:** A community group contends the connector will devastate the African American community by cutting it in half:

**Good:**

**Good:**

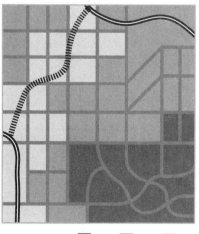

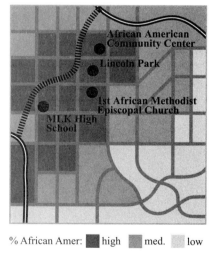

Property Values: ■ high ■ med. ▢ low

% African Amer: ■ high ■ med. ▢ low

28

**Goal:** A historical preservation group shows that historical properties in a historical district will be adversely affected:

**Good:**

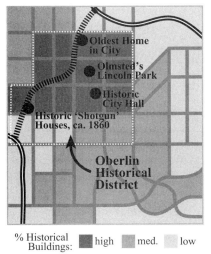

% Historical Buildings: high ■  med. ▨  low ▢

**Goal:** The Oberlin Business Association argues the proposed road will siphon traffic and thus business away from their members:

**Good:**

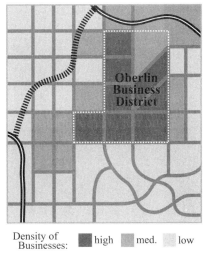

Density of Businesses: high ■  med. ▨  low ▢

**Goal:** An environmental group shows how the proposed connector violates the city's long-standing policy of avoiding road construction in floodplains:

**Goal:** A newspaper story changes the scale to show that the County Chamber of Commerce wants the connector as part of an incentive package to attract a pharmaceutical firm to a suburban development. Most of the employees for the new facility would come from the suburbs south of the city:

**Good:**

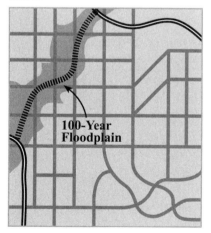

100-Year Floodplain

**Good:**

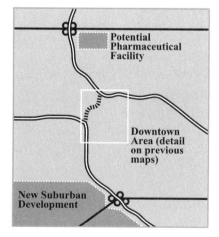

Potential Pharmaceutical Facility

Downtown Area (detail on previous maps)

New Suburban Development

**Different goals produce different maps!**  The eight maps involved in this debate over the location of the connector are all good.  Each is clear.  Each makes its point with accurate data in a way that is easy to read and understand.  What makes the maps different is the different purposes each was designed to serve.  It is this purpose that drove the selection of facts, and these facts that dictated the design and scale.

**Goal:** Due to historical and environmental concerns with the proposed connector, and the embarrassing newspaper article, the planning department is asked to develop two alternative options.  When these alternatives are mapped, they raise additional concerns (and maps):

**Goal:** Alternative B, while more costly than A, is cheaper than C (which passes through property owned by influential developers who don't favor the connector).  B also has a lower environmental impact and does not adversely affect any organized social groups or business interests.

**Good:**

**Good:**

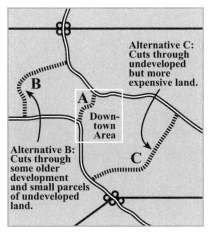

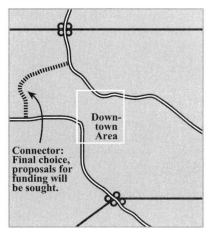

# ②Who is your map for?

**Experts:** Experts know a lot about the subject of the map. Experts are highly motivated and very interested in the facts the map presents. They expect more substance and expect to engage a complex map. An interactive multivariate map for global climate change researchers visually interposes three data variables related to global climate research (hydrological balance, precipitation, and temperature).

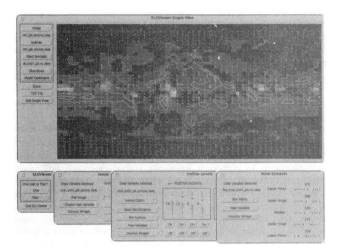

✓ less peripheral information on map explaining content and symbols.

✓ more information, more variables of information, more detail.

✓ follow conventions of experts: consider using a spectral (rainbow) color scheme for ordered data if the user is accustomed to using such colors to show ordered data (spectral color schemes are usually not good for other users).

**Novices:** Novices know less about the map subject and may not be familiar with the way maps are symbolized. They need a map that is more explanatory. Novices may be less motivated than expert users, but they want the map to help them learn something. They expect clarity, and may be put off by a complex map. Farmers need to know about potential drought conditions, but are not trained climatologists. Farmers need a map designed for them.

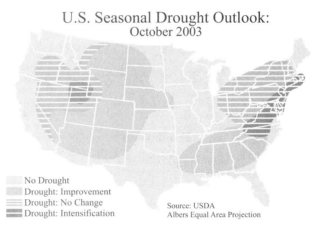

U.S. Seasonal Drought Outlook:
October 2003

No Drought
Drought: Improvement
Drought: No Change
Drought: Intensification

Source: USDA
Albers Equal Area Projection

✓ more peripheral information on map explaining content and symbols.

✓ less information, fewer variables of information, less detail.

✓ follow map design conventions, which enhance comprehension of the map for novice map readers.

# ③What is the final medium?

It is vital to determine the final medium your map will be presented on before making your map. Many maps today are made on computer monitors, but the monitor is not the final medium: paper, a slide to be projected on the wall, projected with a computer, or a poster are common end products. What looks great on your computer will probably not look so great when printed or projected.

## Computer monitor

✓ your map may be designed for the internet. Compared to paper, computer monitors provide less area and lower resolution, suggesting design modifications including larger symbols and type, and fewer data.

## Black and white, on paper

✓ while color is always an option with computers, color may not be an option for your final map: color is expensive to print, and many publications require black-and-white maps.

## Color, on paper

✓ color is created differently on computer monitors and printers. Test your colors on the printer your final map will be printed on, and adjust colors on the computer so they look good when printed.

## Projected

✓ maps projected on a screen using slides or a computer projector require design adjustments, such as larger type and symbols and more intense colors, so that the map is legible from a distance.

## Posters

✓ similar to projected maps, maps in poster form require attention to the typical viewing distance: some posters are to be viewed from a short distance, others from a long distance, and some require both long and short viewing distances.

# Computer monitor: Computer monitor resolution is

about 72 dots per inch (dpi), compared to 1200 or more for many printers. Computer monitors also have limited area: 7 by 9 inches or less if the map is displayed on a web browser. Designing maps for final display on computer monitors must take into account screen resolution and the limited space available. Design a map so that all type and symbols are visible on the screen with no magnification. Also avoid maps that require the viewer to scroll back and forth to see the entire map. Use more than one map if you need more detail.

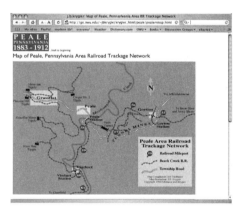

✓ limit map size so that the entire map fits on the screen without scrolling or magnification.

✓ increase type size: 14 point type is the smallest you should use on a monitor.

✓ make point and line symbols 15% larger than those on a paper map.

✓ more distinct patterns: avoid pattern variations that are too fine or detailed.

✓ you may have to limit the amount and complexity of data on your map.

✓ use color: but remember that some monitors cannot display billions of colors; subtle color variations may not be visible on every monitor.

✓ white will be more intense than black; use care when using white to designate no information or as the background.

✓ for the internet, save your map at 72 dpi, and size the map to fit in a browser window (no scrolling).

✓ design your map so it works on different monitors (RGB, LCD).

**Black and white, on paper:** Most maps are created on computer monitors, with less resolution and area than is possible on a piece of paper. When paper is your final medium, design for the paper and not for the monitor. Always check design decisions by printing the map. While all computers offer color, printing with color is not always an option, and many publications require a black-and-white map. Don't despair: much can be done with black and white.

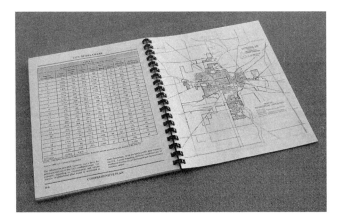

✓ map size should match final paper size, with appropriate margins.

✓ 10-point type works well on a printed map, but you may have to magnify to see it on the computer monitor.

✓ point and line symbols can be smaller and finer on a printed map.

✓ more subtle patterns can be used than on a computer monitor map.

✓ more data and more complex data can be included on a printed map.

✓ substitute a range of greys and black and white for color. Remember that printers cannot display as many greys as you can create on a monitor; subtle variations in greys may not print.

✓ black will be more intense than white; use white to designate no information or the background, dark to designate more important information.

✓ monochrome copiers reproduce grey tones poorly.

✓ very light grey tones may not print.

36

**Color, on paper:** Color on a computer monitor is created in a different manner than color on printers. Select colors on the computer, then print to your final printer. Always design for the final medium: adjust the colors on the monitor so they look best for the printer. The same colors will vary tremendously from printer to printer. Remember that reproducing color is often more expensive than reproducing black and white. Keep in mind that users may reproduce your color map in black and white.

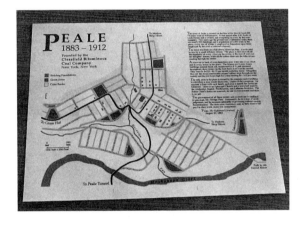

✓ map size should match final paper size, with appropriate margins.

✓ 10-point type works well on a printed map, but you may have to magnify to see it on the computer monitor.

✓ point and line symbols can be smaller and finer on a printed map.

✓ more subtle patterns can be used than on a computer monitor map.

✓ more data and more complex data can be included on a printed map.

✓ use color value (light red vs. dark red) to show differences in amount or importance. Use color hue (blue vs. red) to show differences in kind. Printers cannot display as many colors as you can create on a monitor; subtle variations in colors may not print.

✓ dark colors are more intense than light; use light colors to designate less important information and background, and dark to designate more important information.

✓ never print a color map in black and white; *redesign* for black and white.

**Projected:** Maps on photographic slides or computer monitors can be projected using a slide or computer projector. When projected, white and lighter colors will be more intense, black and darker colors more subdued. Some projectors wash out colors. Projected maps must be designed with the viewing distance in mind (it helps to know the size of the room): a map projected to an audience in a small room can have smaller type and symbols than a map projected in an auditorium. Always check that the map is legible from the back of the room in which the map will be displayed.

✓ greater map size is offset by the increased viewing distance.

✓ increase type size so that smallest type is legible from the back of the room.

✓ increase point and line symbol size to be legible from the back of the room.

✓ more distinct patterns: avoid pattern variations that are too fine or detailed.

✓ you may have to limit the amount and complexity of data on your map.

✓ many projectors wash-out colors, so intensify your colors for projection.

✓ if your map will be projected in a dark room, use black as background, darker colors for less important information, and lighter colors for more important information.

✓ if your map will be projected in a well-lighted room, use white as background, lighter colors for less important information, and darker colors for more important information.

**Posters:** Posters are similar to projected maps, although usually viewed in well-lighted conditions. Unlike projected maps, posters are often not viewed as part of a presentation. Viewers should be able to see the map from afar, then walk up to the map and get more detail. Design the poster so information can be communicated both close and at a distance. The size of poster maps is limited by the largest printer you can use; always check color and resolution of the printer your poster will be printed on.

✓ design map title and mapped area so they are legible from across the room.

✓ the majority of type, point, and line symbols should be slightly larger than on a typical printed map, but not as large as on a monitor or projected map. Design this part of the map so it is legible from an arm's length.

✓ more complex information can be included on a poster map than on a computer monitor or projected map.

✓ follow color conventions for color printed maps.

✓ most posters are viewed in a well-lighted room, so use white as background, lighter colors for less important information, and darker colors for more important information.

# Evaluation

Evaluation should play a role in the entire process. Do I really need a map? Maybe a table or chart or text would be better. If a map is necessary, evaluate it throughout the map-making process. Evaluation during the map-making process is often more important than evaluation of the finished map. While making a map, continually evaluate whether the map is serving its intended goal, meeting the needs of its intended audience, and works well in its final medium. There are different aspects to evaluation.

## Documentation

✓ simply keeping records of what you do while you make your map, for future reference

## Formative Evaluation

✓ continually asking if the map is working while you are making it: iterative process of re-forming the design of the map while making the map

## Impact Evaluation

✓ informal and formal evaluation of the success of the map once it is complete

**Documentation:** What were those six great shades of red I used on that map I made last month? What font did I use on the last poster map I made? How big was the title type? How long did it take me to make that map for the annual report last year? Where did we get that great data set? Was it copyrighted? Who printed that large format map for us last year? How much did it cost to print those color maps in the magazine? Documentation of the myriad details involved in making a map may seem tedious, but can come to save time and effort for future map making, both for yourself and others who may need to make similar maps. Working toward a few general styles that are effective for specific types of commonly produced maps is useful. Documentation of mapped data is vital if the map is to be published.

### Documenting general issues

✓ document your goal for the map.

✓ ...the intended audience, and what you know about them.

✓ ...the final medium, and details about the medium that will affect map design and reproduction.

✓ ...the amount of time it takes to create the map, and any major problems and how you solved them.

✓ keep copies of the map, and information on where it was published or presented.

### Documenting design issues

✓ document specifics of map size, scale, and sketches of layouts.

✓ ...intellectual and visual hierarchies.

✓ ...all data classification and generalization.

✓ ...all sources and the character of map symbols.

### Documenting data issues

✓ document the source of the data, including contact information and copyright information.

✓ ...the age, quality, and any limitations of the data.

✓ ...how the data were processed into a form appropriate for mapping.

✓ ...map projection and coordinate system information.

✓ document details of type size, font, etc.

✓ ...color specifications for all colors used.

✓ ...any design problems encountered, and how you solved them.

✓ ...any software problems encountered, and how you solved them.

**Formative Evaluation:** Formative working evaluation is as simple as asking yourself whether the map is achieving its goals throughout the process of making the map. Formative evaluation implies that you will "re-form" it so it works better, or maybe even dump it! It is never too late to bail if the map is not serving your needs. It is a good idea to ask others to evaluate your map as well: What do you think of those colors? Can you read that type from the back of the room? Does what is most important on the map actually stand out? Simply engaging your mind as you make your map, and being open to critique and change, will lead to a better map.

## Formative Evaluation

✓ is this map doing what I want it to do?

✓ will this map make sense to the audience I envision for it?

✓ how does the map look when printed, projected, or viewed in the final medium, and what changes will make it better?

✓ are the chosen scale, coordinate system, and map projection appropriate?

✓ does the layout of the map and the map legend look good? Could it be adjusted to help make the map look better and be easier to interpret?

✓ does the most important information on the map stand out visually? Does less important information fall into the background?

✓ are data on the map too generalized or too detailed, given the intent of the map?

✓ does the way I classified my facts help to make sense out of them? Would a different classification change the patterns much?

✓ do chosen symbols make sense, and are they legible?

✓ is the type appropriate, legible, and is its size appropriate, given the final medium?

✓ is color use logical (e.g., value for ordered data, hue for qualitative data), and appropriate, and will the chosen colors work in the final medium?

✓ do I want a series of simpler maps, or one more complicated one?

✓ is a handout map needed, if presenting a map on a poster or projected?

42

**Impact Evaluation:** Impact evaluation refers to a range of informal and formal methods for evaluating the finished map. It often consists of your boss or a publisher reviewing the map, or may involve public feedback on the map's efficacy. You should begin any map-making with a clear sense of who may have the final say on the acceptability of your map, and factor in their wants, needs, and requirements at the beginning of the process.

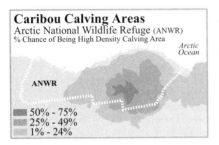

**Caribou Calving Areas**
Arctic National Wildlife Refuge (ANWR)
% Chance of Being High Density Calving Area

Arctic
Ocean

ANWR

50% - 75%
25% - 49%
1% - 24%

Ian Thomas, a contractor for the U.S. Geologic Survey, was fired, allegedly for publicizing maps of caribou calving areas in the ecologically and politically sensitive Arctic National Wildlife Refuge. Thomas and others argue he was fired for publicizing facts that would undermine the push for oil exploration in the Refuge. Others claim the maps were of out-of-date information beyond Thomas' area of expertise and had nothing to do with his firing. In either case, it is obvious that making maps can piss off your boss.

A diagrammatic map of the transport system in Amsterdam, first published in 1989 and similar to the famous London Underground map, won design awards from industrial designers. Alas, it was a flop for the majority of users, and was quickly replaced by a more conventional map.

He picked out, with the help of the girls, where they were on the map flowing across his lap board. He had not known until recently that humans knew about maps. It had given him a twinge of happy homesickness the first time he had grokked a human map. It was static and dead compared with maps used by his people – but it was a map. Even human maps were Martian in essence – he liked them.

Robert Heinlein. *Stranger in a Strange Land.* (1961)

Boundary, n. In political geography, an imaginary line between two nations, separating the imaginary rights of the one from the imaginary rights of the other.

Ambrose Bierce. *The Devil's Dictionary.* (1911)

The most remarkable escape story of all concerns Havildar Manbahadur Rai ... He escaped from a Japanese prison camp in southern Burma and in five months walked 600 miles until at last he reached the safety of his own lines. Interrogated by British intelligence about his remarkable feat, Manbahadur told them that ... he had a map, which before his capture had been given to him by a British soldier in exchange for his cap badge. He produced the much creased and soiled map. The intelligence officers stared at it in awe. It was a street map of London.

Byron Farwell. *The Gurkhas.* (1984)

Maps encourage boldness. They're like cryptic love letters. They make anything seem possible.

Mark Jenkins. *To Timbuktu.* (1997)

44

# more information...

Three good books about maps, GIS, and local political action are Doug Aberley's *Boundaries of Home: Mapping for Local Empowerment,* (New Society Publishers, 1993); William Craig et al., *Community Participation and Geographic Information Systems* (Taylor and Francis, 2002); and Kim English and Laura Feaster, *Community Geography* (ESRI Press, 2003).

Brian Harley's excellent essays on maps, history, and power are collected as *The New Nature of Maps: Essays in the History of Cartography* (Johns Hopkins University Press, 2002).

Mark Monmonier's thought-provoking books are packed with stories about how maps work in the world (all University of Chicago Press unless noted): *Maps with the News: The Development of American Journalistic Cartography* (1989), *Drawing the Line: Tales of Maps and Cartocontroversy* (Henry Holt, 1994), *Cartographies of Danger: Mapping Hazards in America* (1997), *Air Apparent: How Meteorologists Learned to Map, Predict, and Dramatize Weather* (1999), *Bushmanders and Bullwinkles: How Politicians Manipulate Electronic Maps and Census Data to Win Elections* (2001), and *Spying with Maps: Surveillance Technologies and the Future of Privacy* (2002). *Mapping it Out: Expository Cartography for the Humanities and Social Sciences* (1993) contains an excellent overview of cartography and map design.

**Sources:** The multivariate map opening the chapter is from Stuart Smith et al., "Global Geometric, Sound and Color Controls for Iconographic Displays of Scientific Data" (*Proceedings of SPIE,* vol. 1459, 1991). The interactive "expert" map (p. 32) is discussed in David DiBiase, "Multivariate Display of Geographic Data" in A. MacEachren and D.R. Fraser Taylor, *Visualization in Modern Cartography* (Pergamon, 1994). The caribou calving map (p. 43) is re-created from data and maps at Ian Thomas' web site. The bit from the Amsterdam transport map (p. 43) is from H. van der Kooy's *Informatie Openbaar Vervoer in Amsterdam* (Gemeente Vervoerbedrijf, 1989) and story from Paul Mijksennar's *Visual Function* (Princeton University Press, 1997).

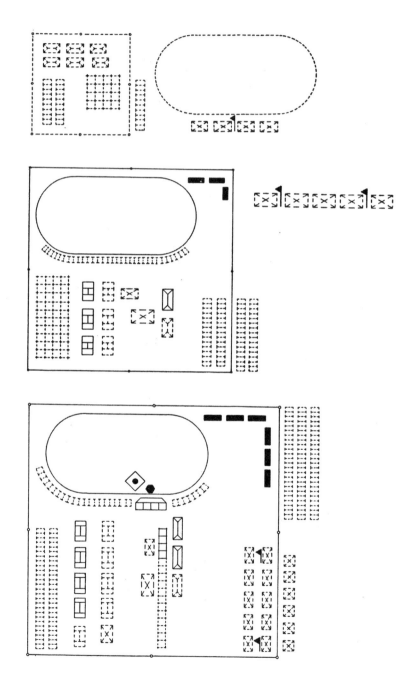

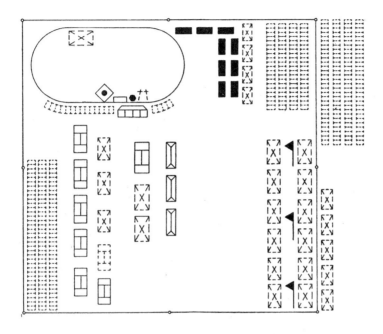

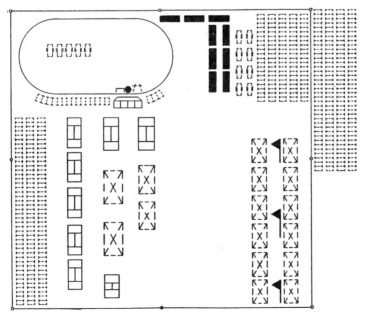

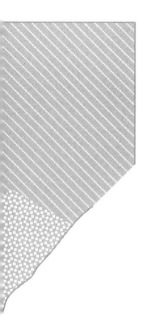

*What do maps show us?*

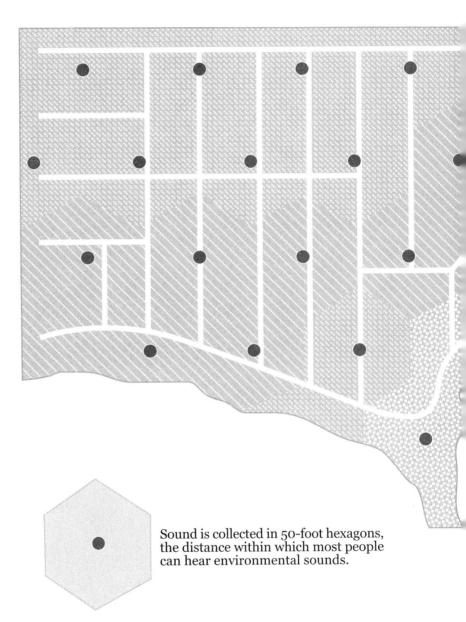

Sound is collected in 50-foot hexagons, the distance within which most people can hear environmental sounds.

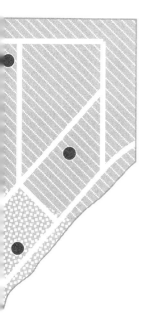

# Mappable Data

***Maps show us data.*** This maps *shows* environmental sound in part of Vancouver, British Columbia.  Recording stations collect sounds at different locations, and the map groups the sounds into three categories:

Low frequency & high variety
   Less vehicle noise, more natural and human sounds
High frequency & high variety
   More vehicle noise, less natural and human sounds
High frequency & low variety
   Major highway, very high vehicle noise

You can map just about *any* data you can collect from the environment.

# Mappable Data

Between deciding what you want to map and mapping it lies the important task of acquiring mappable data. The data you need to make your map may or may not be easily available, or in a usable form. When making maps, budget both time and money for data, and leave sufficient time for acquiring and processing the data. Understanding mappable data involves eight issues:

## Phenomena & data

Maps display data, and data distill human and physical phenomena. It is vital to distinguish data from phenomena when making maps.

## Data layers

Some data layers provide the backdrop for other data on a map. Data may come from different sources, and must be processed to work together.

## Getting data

We acquire data directly from the environment, or use already collected data from an existing source.

## Data organization

Mappable data are organized as either raster or vector format. Your goals for your map and the software you are using determine which format you should use.

## Quantifying data

Data can be qualitative or quantitative, and there are several different kinds of each. The level of quantification will shape how you symbolize your data.

## Transforming data

Common processing of data makes them more mappable, including averages, densities, and ratios.

## Data accuracy

Accuracy is complicated! There are many aspects of data accuracy you must assess when working with mappable data.

## Digital data and GIS

Digital data require understanding issues of metadata and copyright.

# 1 Phenomena & data

Phenomena are all the stuff out in the real world. Data capture specific phenomena. Keep in mind that maps do not directly display phenomena: they display data. Some maps are designed to mimic phenomena, and other maps are designed to mimic the data. The surface map (below) shows temperature in a manner that mimics the phenomena. The population map (bottom) reveals more about the data (one value for each county) than the phenomena (people are not evenly spaced in a county).

**Phenomenon:** temperature is found everywhere in varying degree.

**Data:** known temperatures at a few locations.

**Map:** shows semicontinuous change in temperatures, extrapolated from known temperatures at a few locations.

**Comment:** map suggests the phenomenon of temperature.

**Phenomenon:** people in Utah (low to high).

**Data:** U.S. Census count of how many people are in each county.

**Map:** map may suggest an even spread of people throughout each county.

**Comment:** map displays the character of the data (total number *by county*) rather than the phenomenon (where people actually live).

54

# Data layers

Some data layers provide context and reference for other data layers on a map. Road and municipal boundary data may serve as the backdrop on a map showing variations in zoning in a town (created for a meeting about proposed zoning changes). In this case the zoning data are of most importance, and the road and boundary data are included to help understand the zoning data. Data layers often come from different sources, and coordinate systems and projection may have to be adjusted for the data to work together.

**Poor data choice & design:**

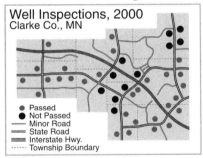

Well Inspections, 2000
Clarke Co., MN

● Passed
● Not Passed
— Minor Road
═ State Road
▬ Interstate Hwy.
⋯⋯ Township Boundary

**Map intent:** to show well inspections for 2000 and whether they passed inspection for a county commissioners' meeting.

**Vital data:** well locations and status of the well inspections.

**Background data:** roads, county outline, and township boundaries.

**Problem:** background data visually over-whelm the vital well data; too many roads, and the river (which seems to be related to well failure) is missing.

**Good data choice & design:**

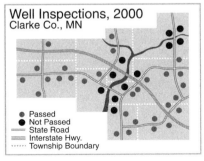

Well Inspections, 2000
Clarke Co., MN

● Passed
● Not Passed
═ State Road
▬ Interstate Hwy.
⋯⋯ Township Boundary

**Data and design adjustments:** the map needs to effectively show the well inspection data. Data layers which *do not* help understand the well data, such as minor roads, can be removed. Data which *do* help to understand the well inspection data, such as rivers in this case, need to be included. Other data can be redesigned to be less noticeable, such as the township boundaries and roads.

# ③ Getting data

Data can be acquired directly from the environment (primary sources) or from others who have already collected the data (secondary sources). Most map makers use secondary sources, but various methods make it possible for anyone to collect mappable data. Primary and secondary data are often combined on maps, and care must be taken that the data have the same coordinate system and projection.

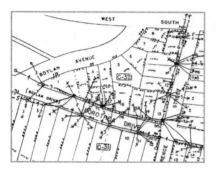

**Data compilation on existing maps:** An easy and inexpensive way to collect data is simply to record them on an existing map. Researchers used a city property map to record the location of electrical poles and powerlines. Such data can be digtized or scanned and used with existing layers of data in GIS.

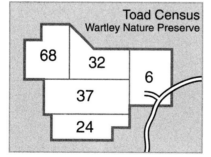

**Census:** A census involves defining an area, and counting how many of some phenomena are in that area. A count of toads in five parts of a nature preserve is a census. A census reveals how many of something is in an area, but not where they are in the area. Such census data can be added to existing GIS data layers that contain your census boundaries. A national census, like the U.S. Census, uses many census data collection techniques.

**Collecting data at addresses:** Collecting data by street address is also easy and inexpensive. Researchers recorded 20 different facts about the exterior of houses (part of the data collection sheet is shown). Such data can be "address matched" to existing GIS data, including ownership and tax information.

**Global Positioning Systems (GPS):** a means of collecting primary data. A series of satellites relay signals which a GPS receiver uses to determine the location of the device. Simple and cheap GPS receivers provide location and elevation. More sophisticated devices allow you to append attribute information (what is at the location recorded by the GPS device) and export the data so that it can be easily mapped in GIS.

**Characteristics:** location of points, lines, and areas (vector data), sometimes with added attribute data (oak tree at point A).

**Remote sensing imagery:** photographic or digital images of the earth, taken from airplanes or satellites (remote sensors). Remote sensing imagery is available at different scales (earth to your street). Remotely sensed imagery can serve as the source of mappable data: roads can be traced or vegetation types delineated. Remotely sensed imagery is often combined with other map data in GIS.

**Characteristics:** a very fine grid of varying values representing variations in energy reflected from the earth (raster data).

**Road data:** from United States Geological Survey (USGS)

**Air imagery:** from private areal photography company

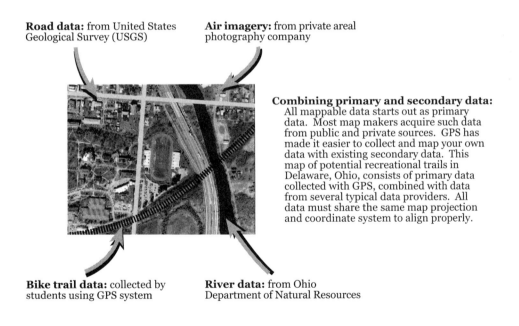

**Combining primary and secondary data:** All mappable data starts out as primary data. Most map makers acquire such data from public and private sources. GPS has made it easier to collect and map your own data with existing secondary data. This map of potential recreational trails in Delaware, Ohio, consists of primary data collected with GPS, combined with data from several typical data providers. All data must share the same map projection and coordinate system to align properly.

**Bike trail data:** collected by students using GPS system

**River data:** from Ohio Department of Natural Resources

# **4** Data organization

There are two basic ways that geographic data are organized: vector or raster.  Vector data consists of points, which can be connected into lines, or areas.  Raster data consists of a grid of cells, each with a particular value or values.  Vector and raster data can be used together when making maps.

## *vector*

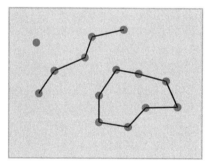

**Characteristics:** vector data consist of located points, lines (a connected series of points), and areas (a closed, connected series of points). Attribute information can be appended to a point, line, or area. A line representing a road could include attributes such as name, width, surface, etc.

**Sources and use:** GPS devices collect vector data; many public (USGS, Census TIGER), and private sources of mappable data provide data in vector format.  Most GIS software uses vector data.

## *raster*

**Characteristics:** raster data consist of a grid with values associated with each cell in the grid.  Higher resolution raster files have smaller cells.  Remotely sensed imagery is in raster format: each cell represents a level of energy reflected or radiated from the earth in the area covered by the cell.  Raster data can have points (one cell), lines (a series of adjacent cells), or areas (a closed series of adjacent cells). Raster data can also include attributes.

**Sources and use:** The most common mappable raster data is satellite and areal imagery, available from public and private sources.  Most GIS software allows you to use raster and vector data together.

# 5 Quantifying data

At a simple level, mappable data is either qualitative - differences in kind - or quantitative - differences in amount. Such variations in data guide map symbolization.

## *qualitative*

**Differences in kind:** also called nominal data

**Examples:**

✓ house and business locations
✓ rivers and lakes
✓ electorial college wins, by state (Democrats or Republicans)
✓ dominant race in a block-by-block map of a town
✓ location of different bird species seen in a nature preserve

**Symbolization:** shown with symbols, pictographs, or icons; or with differences in color hue (red, green, blue), as such colors are different in kind, like the data

## *quantitative*

**Differences in amount:** includes ordinal, interval, and ratio data

**Examples:**

✓ estimated number of same-sex couples, living together, in the U.S. (by county)
✓ total number of Hispanics in a block-by-block map of a town
✓ number of a loggerhead shrikes counted in a nature preserve

**Symbolization:** shown with differences in color value (dark red, red, light red), as such colors are different in amount, like the data

**Levels of quantification:**

✓ **ordinal:** order with no measurable difference between values: low, medium, and high risk zones

✓ **interval:** measurable difference between values, but no absolute zero value: temperature Fahrenheit (30° is not half as warm as 60°)

✓ **ratio**: measurable difference between values, with an absolute zero value: total population

# ⑥ Transforming data

Data are often transformed before being mapped. Averages, densities, and ratios can be derived from total numbers.

## total numbers

**Characteristics:** The total number of some phenomenon associated with a point, line, or area.

**Examples:**

- ✓ amount of pesticide in a well
- ✓ 24-hour traffic count on road
- ✓ road-kill collected in a county

**Symbolization:** Variation in point size or line width. Represent whole numbers in areas with a scaled symbol for each area.

## densities

**Characteristics:** The number of some phenomenon per unit area. Appropriate for data in areas. Divide the total number of people in a county by the square miles of the county.

**Examples:**

- ✓ population per square mile in a county
- ✓ doctors per square km in a country
- ✓ adult bookstores per square mile in cities

**Symbolization:** Variation in lightness and darkness in the areas.

## averages

**Characteristics:** Add all values together and divide by the number of values in the data set. Can be associated with points, lines, and areas.

**Examples:**

- ✓ average monthly rainfall at a weather station
- ✓ average monthly flow in a river
- ✓ average age of gang murder victims in U.S. cities

**Symbolization:** Variation in point size or line width. Variation in lightness and darkness in areas.

## rates

**Characteristics:** The number of some phenomenon per unit time. Can be associated with points, lines, and areas.

**Examples:**

- ✓ gallons per minute from a well
- ✓ cars per hour along a road
- ✓ murders per day in major cities

**Symbolization:** Variation in point size or line width. Variation in color lightness or darkness in areas.

# **Data accuracy**

There are many types of accuracy associated with data and maps. One approach to accuracy is to ask a series of questions about your data.

## *are facts accurate?*

**Example:** Detailed U.S. Geologic Survey maps of the U.S. do not include sensitive military and archaeological sites.

**Example:** A map in Microsoft Windows 95 showed Kashmir as a disputed territory. India claims Kashmir, and halted government purchases of Windows 95 software.

## *does detail vary across the data?*

**Example:** A sand and gravel database of Wisconsin combines highly detailed data from some counties with crude data from other counties.

**Example:** Internet map sites of roads vary in accuracy across the U.S.

## *are data from a trust-worthy source?*

**Example:** The U.S. Geologic Survey has published standards for data accuracy and quality.

**Example:** Data from an unknown source, with no information about when the data were collected and no information about accuracy should be used with caution.

## *are things where they should be?*

**Example:** All roads in the Delaware County, Ohio, database are within 5 feet of their actual location.

**Example:** The National Imagery & Mapping Agency had the Chinese Embassy in the wrong location on a map of Belgrade, Yugoslavia. A NATO jet accidentally bombed the Embassy, killing 3 and injuring 20.

## *when were the data collected?*

**Example:** U.S. Census data are collected every ten years: a map made in 2009 using 2000 U.S. census data will have nine-year-old data.

**Examples:** When were roads last updated on the internet road map site?

## *are data appropriate for the map goal?*

**Example:** A tourist map promoting Brazil may not require highly detailed geological data.

**Example:** Official maps of the Arctic National Wildlife Refuge may not show caribou calving areas.

# Digital data and GIS

Widespread use of Geographic Information Systems and development of extensive databases of digital GIS data require an understanding of metadata and copyright.

## *metadata*

**Metadata** are data about data. Dependable digital geographic data will include detailed information including:

- ✓ *identification information:* general description of the data
- ✓ *data quality information:* in terms of data quality standards
- ✓ *spatial data organization information:* how spatial information in the data is represented
- ✓ *spatial reference information:* coordinate and projection information
- ✓ *entity and attribute information:* map data and associated attributes
- ✓ *distribution information:* data creator, distributor, and use policy
- ✓ *metadata reference information:* metadata creator
- ✓ *citation information:* how to cite information when used
- ✓ *temporal information:* when data was collected, updated
- ✓ *contact information:* how to contact data creator

Source: Longley et al., *Geographic Information Systems and Science,* 2001

**Metadata** standards in the U.S. have been set by the Federal Geographic Data Committee (FGDC, www.fgdc.gov). Geographic data providers often follow such guidelines. For example, see the extensive geographic data available at www.geographynetwork.com, all of which are provided with metadata that meet FGDC standards.

# copyright

**Copyright** is a form of protection provided by the laws of a country to the authors of original works of authorship. In the U.S., copy "rights" include:

- ✓ reproduction of copies of the original copyrighted work
- ✓ preparation of derivative works based on original copyrighted work
- ✓ distribution/sale/transfer of ownership of original copyrighted work

Maps, globes and charts are covered under U.S. copyright law (17 U.S.C., sec. 101), and you may officially register a map or simply include a copyright statement on it. Court cases have established that "facts" may **not** be copyrighted, but a particular representation of those facts can, as long as the representation includes an "appreciable" amount of original material. Thus someone can make a map based on the data included on your copyrighted map, but they cannot photographically reproduce it.

Developers of GIS data argue that their data are a ***representation*** of facts, and **not** facts themselves, and thus such data can be copyrighted. Licensing GIS data is another way to control access to data you have collected.

As a data user, always seek information (in the metadata) about copyright and licensing and other restrictions on map and data use. As a rule of thumb, federal government maps and data in the U.S. are free of copyright or licensing restrictions, but all other maps and data, including data created by states, local governments, planning agencies, and private firms, may be copyrighted.

So Joshua said to the Israelites: "How long will you wait before you begin to take possession of the land that the LORD, the God of your fathers, has given you? Appoint three men from each tribe. I will send them out to make a survey of the land and to write a description of it, according to the inheritance of each. Then they will return to me. You are to divide the land into seven parts. Judah is to remain in its territory on the south and the house of Joseph in its territory on the north. After you have written descriptions of the seven parts of the land, bring them here to me and I will cast lots for you in the presence of the LORD our God. The Levites, however, do not get a portion among you, because the priestly service of the LORD is their inheritance. And Gad, Reuben and the half-tribe of Manasseh have already received their inheritance on the east side of the Jordan. Moses the servant of the LORD gave it to them." As the men started on their way to map out the land, Joshua instructed them, "Go and make a survey of the land and write a description of it. Then return to me, and I will cast lots for you here at Shiloh in the presence of the LORD." So the men left and went through the land. They wrote its description on a scroll, town by town, in seven parts, and returned to Joshua in the camp at Shiloh.

*Joshua 18, 3-9*

Our maps are candid charts of desolation.

Ivor Brown. *The Moorland Map.* (1943)

66

# more information...

Just about any GIS textbook will discuss mappable data issues in much detail. A good place to start is Paul Longley et al., *Geographic Information: Systems and Science*, 2nd ed. (Wiley, 2005). Michael Zeiler's *Modeling Our World: The ESRI Guide to Geodatabase Design* (ESRI Press, 1999) is a great introduction to how geodatabases work.

A good start with copyright issues and mapping are a pair of articles by Mark Monmonier: "Map Traps: The Changing Landscape of Cartographic Copyright" (*Mercator's World,* July/August 2001), and "Originality Bites: Copyright Registration and Map History" (*Mercator's World,* September/October 2001). An excellent overview of copyright and licensing issues related to GIS data is William Holland, "Copyright, Licensing and Cost Recovery for Geographic and Land Information Systems Data: A Legal, Economic and Policy Analysis" (*Proceedings Of The Conference On Law And Information Policy For Spatial Databases,* National Center for Geographic Information and Analysis, 1994).

For a ghostly tale involving inaccurate, out-of-date maps read A.N.L. Munby's "An Encounter in the Mist" (reprinted in *The Oxford Book of English Ghost Stories,* Oxford University Press, 1987).

**Sources:** The sound map of Vancouver, British Columbia (pp. 48-51) is redrawn from J. Douglas Porteous and Jane Mastin, "Soundscape" (*Journal of Architectural Planning Research,* vol. 2, 1985). Delaware County, Ohio, air imagery (pp. 59-60) courtesy of the DALIS Project, Delaware County, Ohio.

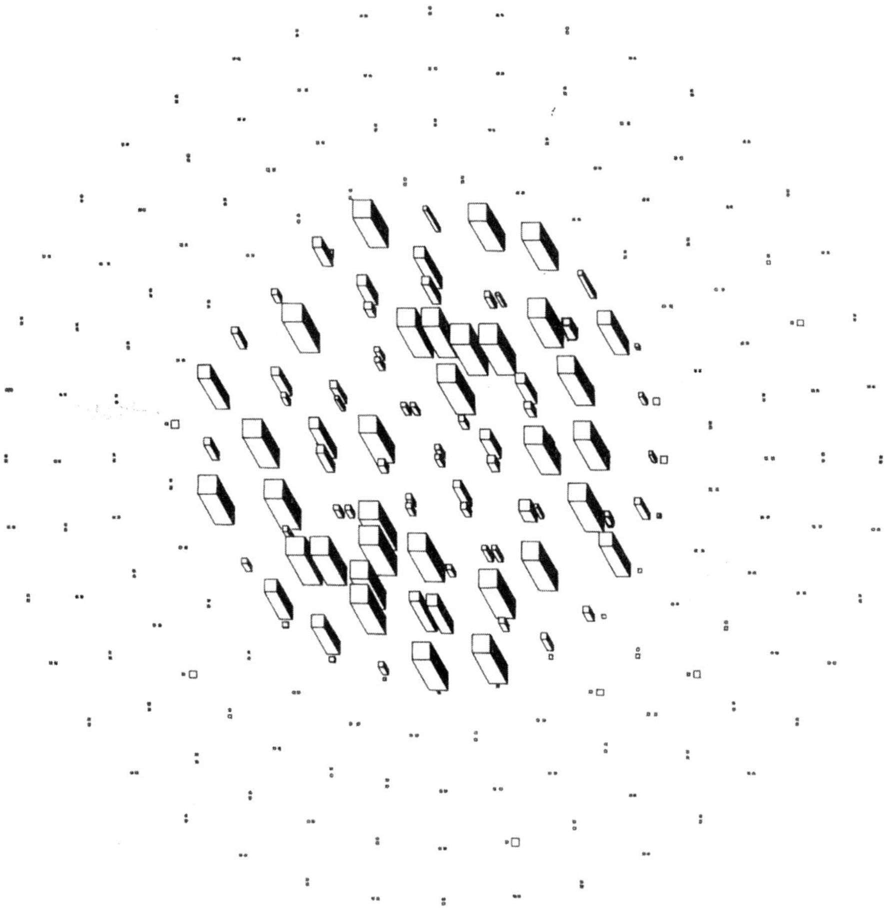

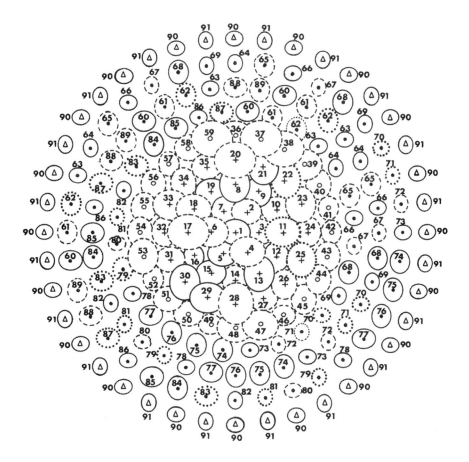

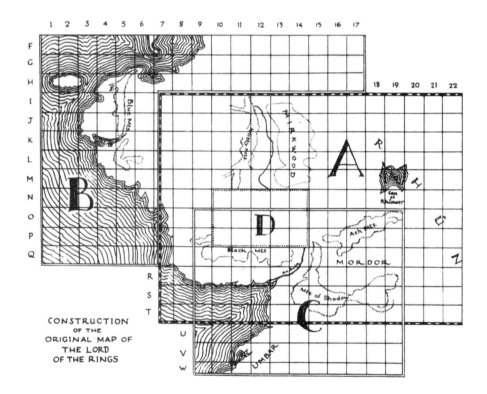

CONSTRUCTION
OF THE
ORIGINAL MAP OF
THE LORD
OF THE RINGS

*How is it made?*

J.R.R. Tolkien's son, Christopher, made this diagram of the structure of his father's continuously evolving map of Middle Earth. Sheet **A** was the first to be drawn, and includes the lands mapped for the elder Tolkein's first book, *The Hobbit.*

The extension of the map (**B**) added the coastlands necessary for contextualizing Middle Earth in the "background" story of *The Silmarillon,* which the elder Tolkien had been working on for years.

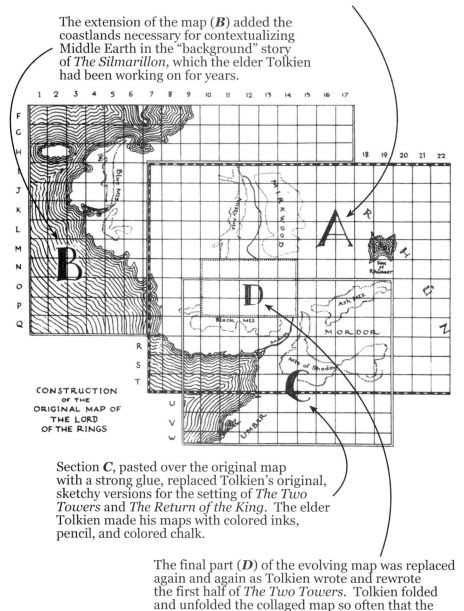

CONSTRUCTION
OF THE
ORIGINAL MAP OF
THE LORD
OF THE RINGS

Section **C**, pasted over the original map with a strong glue, replaced Tolkien's original, sketchy versions for the setting of *The Two Towers* and *The Return of the King.* The elder Tolkien made his maps with colored inks, pencil, and colored chalk.

The final part (**D**) of the evolving map was replaced again and again as Tolkien wrote and rewrote the first half of *The Two Towers.* Tolkien folded and unfolded the collaged map so often that the crease marks are hard to distinguish from the joins between the pieces.

# Map-Making Tools

***The reason a map is being made suggests appropriate tools.***
Even if J.R.R. Tolkien knew how to use fancy GIS software, it would not
have been a suitable tool for the map he developed in tandem with his
*Lord of the Rings* trilogy. The map of Middle Earth was iteratively
developed as Tolkien wrote his books, and consists of various pieces
superimposed and glued together over time. Pencils, ink, chalk, paper,
and glue are great map-making and thinking tools. So too can be internet
mapping sites, and GIS software. Choose appropriate tools based on
what you need to do.

# Map-Making Tools

Making maps requires tools, ranging from a pencil to sophisticated GIS software. Choosing which tool is appropriate depends on your reason for making the map. To work with the chosen tools, your data may have to be transformed. Map-making tools vary in their ability to control map design. It is often necessary to use different map-making tools together. Extensive resources exist to help you to learn how to use the tools you choose.

## 1 Making maps without computers
Map making without a computer is possible.

## 2 Making maps on the internet
The internet has democratized map making.

## 3 Making maps with GIS
Sophisticated spatial analysis and map making are aided by GIS.

## 4 Graphic Design Tools
Graphic design software helps you to create refined maps.

# 1 Making maps without computers

You certainly don't need a computer to make maps. Indeed, map making with pencils and paper is appropriate, inexpensive, and effective in *many* instances. A sketch map made with pencils and paper may be your final map, or it may be a vital step in the process of producing a map with other tools.

**Win the election!** Volunteers for the Democratic presidential candidate go door to door in a key city in a swing state. Using a map printed from an internet site and a pen, they mark their opponent's supporters as an **X** and their supporters as an **O**. A filled **O** means a supporter who may not vote (such as an elderly person with no transportation). These people will be called on election day to urge them to vote and to offer free transport to the polls.

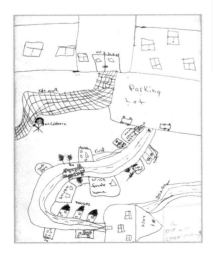

**Sketch maps:** To learn something about the effect of riding the bus on middle schoolers' home ranges, G. T. Wells compared sketch maps made by walkers and riders. This map was made by Chris, who rode bus #71. In this situation, nothing more than a sketch map with pencil and pen is necessary.

# **2** Making maps on the internet

The internet has democratized map making. Anyone with access to the internet can make maps on sophisticated interactive mapping sites. Such sites are free and easy to use but limited in functions, compared to mapping software you can purchase. Internet mapping sites are increasingly sophisticated, with more mapping functions and more control over map design.

## *use the internet for...*

**Getting mappable data:** The internet provides an excellent way to locate and acquire mappable data. Some data are free, some are available for a fee. The best mappable data sites include meta-data (data describing the data) and the capability to immediately download data.

**The Geography Network** is a collaboration between private and public sector data and map professionals, and provides access to a diversity of data, maps, and map-related services. Metadata and searching functions make data access easy.

**Ordering paper maps:** You can order many things on the internet, including cheese, ladybugs, fine cutlery, tasty meats, and honest-to-goodness paper maps. Such maps are an excellent source for making new maps.

**The United States Geologic Survey (USGS) Map Locator** helps locate and order paper maps from their highly detailed topographic map series.

**Perry-Castañeda Library
Map Collection**

---

**Online Maps of Current Interest**

- Afghanistan Maps including many links to maps on other web sites
- India Maps, Pakistan Maps and Kashmir Maps
- Iraq Maps
- Israel Maps with Gaza and West Bank
- Philippines Maps

**Online Maps of General Interest**

- Maps of The World
- Maps of Africa
- Maps of The Americas
- Maps of Asia
- Maps of Australia and the Pacific
- Maps of Europe
- Maps of The Middle East

**Map Snapshots:** Static digital copies of existing paper maps as well as digital images of computer-generated maps are available from map libraries, government sources, and some commercial sites. Such maps are useful as sources for making maps. Most map snapshots are in raster format.

**The Perry-Castañeda Library Map Collection** at the University of Texas has thousands of copyright-free digital maps available.

**Locations and directions:** Many internet sites allow you to map locations and routes. Such sites replace the need to sketch directions by hand, or plot them on a paper map. The maps at these sites may also be useful as a data source for making maps, but be careful about copyright and the skimpy documentation on data quality and sources.

**MapQuest, MapsOnUs, Expedia, Yahoo! Maps, and Google Maps** are but a few of the many free commercial sites where you can make maps of locations and routes.

**Map Projections:** A few internet sites allow you to generate maps of the world in different map projections. Basic map design options are available, and the map you generate can be saved.

**Design Map from Australia's Charles Sturt University** is an internet site which provides tools for making a dozen different map projections of the world or parts of it. The results can be saved as a raster file.

# use the internet for...

**Extensive and detailed sets of human and environmental data** can be mapped using internet mapping sites. Some of the most important data are provided by government agencies.

**The U.S. Census American Factfinder** can be used to download mappable data and to create maps of U.S. Census data at different scales: national, state, county, city, as well as the finer Census tracts, block groups, and blocks. The internet site is designed like a simple GIS, and allows for some map design choices.

**The U.S. Environmental Protection Agency's Enviromapper** site provides extensive information on the location of toxic releases down to a neighborhood scale. The site works like a simple GIS, allowing you to access detailed information about particular toxic emission sites.

You could easily make and compare maps of your neighborhood containing both U.S. Census data (American Factfinder) and toxic release data (Enviromapper). The results are both raster files, and cannot easily be combined into a single map.

78

**Perform basic GIS analysis:** The future of GIS may be on the web: you access (and pay for) use of GIS based on time and analytical demands. Basic GIS functions such as query and buffering are available on some internet sites.

**The Delaware (Ohio) Area Land Information System (DALIS)** was one of the first regional internet-based GIS sites. Many focus on property parcels and property information at the county level, and also include data on roads, rivers, political boundaries, and aerial imagery.

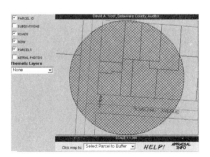

**The DALIS site** allows you to search for a particular property in the county, then create a "buffer" around that property. You may ask the site, for example, to show all properties within 100 feet of John Smith's property. Such functions are very useful if Mr. Smith wants a liquor license for his restaurant and needs to inform his neighbors.

79

# Making maps with GIS

Geographic Information Systems (GIS) software is not only for making maps. GIS assists in the manipulation and analysis of geographic data including *queries* (show all parcels owned by the city), *buffers* (show all roads within 100 feet of wetlands), and *overlay* (combine a map of soils with a map of wetlands, and show only the soils that underlie wetlands). In most cases, GIS users interact with maps while engaged in GIS analysis, and the outcome of the analysis is a map. Because of this, most GIS software has map-making and map design capabilities.

**Recreation Trails in Delaware, Ohio:** A collaboration between students, planners, and community members uses GIS to develop a plan for trail development in the county. GIS data (roads, property parcels, rail roads, etc.) are acquired. *Queries* to the parcels layer produces a map of all schools (flagged symbol) and parks (grey areas). Potential trails should connect schools and parks, and this map helps the collaborators generate potential trail corridors that do just that.

**Abandoned railroad grades** can be converted to trails. Residential property owners adjacent to rail grades sometimes object to rail-to-trail conversions. A 600 foot *buffer* is generated around the rail grades. *Overlay* the buffers on the parcels layer to determine the percent of adjacent properties that are residential, commercial, and industrial. Rail grades with more adjacent residential property may arouse more concern in the city. This map helps to anticipate potential conflict over rail-to-trail development.

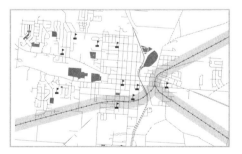

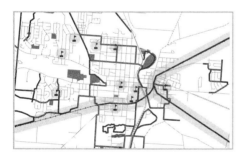

Collaborators use GIS analytical tools to generate potential trails that maximize connectivity between schools and parks but minimize potential conflict. To this "working map" in GIS is added potential trail routes, collected with GPS. This map embodies the process of using GIS and maps to generate a speculative network of recreational trails.

It is time to go public with the trails map, but adjustments need to be made. Schools and parks are labeled, and roads added for reference. Contact with the railroad reveals they are interested in selling the rail grade to the city, so that potential trail is left on the map. Another potential trail is of lower priority, and is adjacent to property owners who will probably object, so it is removed from the map. The map is designed to look professional, suggesting that the process of generating potential trails was professional.

# Graphic Design Tools

Internet mapping tools offer limited control over map design, and the results are low-resolution raster files. GIS mapping tools offer more control over map design, and higher-resolution output, but still may not meet the needs of those who produce high-quality maps for publication. Software designed for graphic design offers extensive control over text, color, and all point, line, and area symbols on a map. While not designed with map making in mind, graphic design software can import raster maps or maps generated by GIS, re-create and redesign them, and generate files that can be professionally printed.

**Enviromapper** internet site generates maps that locate sites in the EPA's Toxic Release Inventory. Great information, but the map is inadequate for a project focused on mapping toxins by zip code in Columbus, Ohio. The toxic release sites and schools are not easy to see, and there are too many roads. Save the map from the internet as a raster (.jpg) file and import it into graphic design software.

**The re-created and redesigned toxins map** better suits the needs of the community group interested in drawing attention to toxic release sites near schools. The toxic site and school symbols are re-created much larger, distracting details are left off the map, and the zip code area is easier to see. Graphic design software provides extensive control over map design and helps create more effective maps.

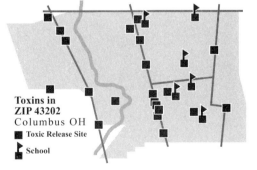

**Toxins in ZIP 43202**
Columbus OH

■ Toxic Release Site

▶ School

# *map-making tools work together*

Strive to envision the best map possible based on what you need your map to do, and harness different software to make your vision work:

**Goal:** Help explain rural to urban migration in the U.S. over time, for a web site devoted to geographic education. A good choice is an animated choropleth map of population change in each U.S. state from 1900 to 2000, using U.S. Census data.

**Data source:** U.S. Census Bureau, historical Census data (text file) accessed via the Census internet site using an *internet browser.*

**Data processing:** use a *spreadsheet* to import, clean, and reformat the text data into a database (DBF format) file.

**GIS:** import the DBF file into the *GIS software* and join to an existing map of each state. Calculate percent change from decade to decade (1900-1910, 1910-1920, etc.), ending with ten sets of percent change data. Classify the ten sets of data: select one classification scheme that fits all ten maps (so they are comparable when animated). Select appropriate colors and generate the ten maps. Export each map and legend as a raster (.jpg) file.

**Image processing software:** import the .jpg files into *image processing software,* adjust the size and colors of each file, and save in another raster format that is easy to animate on the internet (.gif).

**Animation software:** import the ten .gif files into *GIF Animation software* and create a single animation file.

**Web page:** create a web page using *web scripting software,* and embed the animated maps; add explanatory information, and upload to an internet server.

*No single software package can do everything!*

"When John showed me his provisional maps, trying to give me an idea of how many miles his [exploratory] party would cover each day, I thought I might set up a duplicate map, drawing to scale all the country between St. Louis and South Pass.  We know what day he expects to leave St. Louis; I could start him off on that date, drawing a line for the extent of ground he might cover in that day, then put a red dot where they would camp for the night and build their fires....  I had thought I would read the available accounts of what the country is like for each day's march: what wild life is there, what kind of animals they will be shooting for their food.  I would draw my map to show the plains and forests and rivers and mountains, and dotted through the map, I would paint in what I imagined the country looked like, with a small field of wild flowers, a patch of pine forests, a few buffaloes roaming across a plain ..."

"You've set quite a task for yourself," said her father, amused.

"But don't you see, Father," she cried passionately, "in that way he never leaves me, I'll never be alone, I'll be with him on the trail every hour."

> Irving Stone. *Immortal Wife: The Biographical Novel of Jessie Benton Fremont.* (1944)

...The Indians are very expert in delineating countries upon bark, with wood coal mixed with bear's grease, and which even the women do with great precision....

> John Long. *Voyages and Travels.* (1791)

84

# more information...

The second half of David Greenhood's *Mapping* (University of Chicago Press, 1944, reissue 1964) is called "Making Your Own" and is all about making maps without computers. The fifth chapter of Doug Aberley's *Boundaries of Home: Mapping for Local Empowerment* (New Society Publishers, 1993) is "How to Map Your Bioregion: A Primer for Community Activists." It adds tools like photocopiers to the tool kit Greenhood recommends.

Heaps of manuals and guides explain how to use GIS software, and internet mapping sites are usually easy to use without much guidance.

For an excellent overview of changing map-making tools, see Mark Monmonier's *Technological Transition in Cartography* (University of Wisconsin Press, 1985).

For a discussion of the development of internet mapping and its cultural and social context, see Jeremy Crampton's *The Political Mapping of Cyberspace* (University of Chicago Press, 2004).

**Sources:** The Tolkien map (pp. 70, 72) is from J.R.R. and C. Tolkien, *The History of Middle Earth, vol. 3* (Houghton Mifflin & HarperCollins Publishers, 1989). Internet maps on p. 76 courtesy of the Geography Network and the U.S. Geologic Survey Map Store. Internet maps on p. 77 courtesy of the Perry-Castañeda Library Map Collection, MapQuest.com, and Design Map (Charles Sturt University). Internet maps on p. 78 courtesy of the U.S. Census Bureau's American Factfinder and the U.S. Environmental Protection Agency's Enviromapper (also p. 82). Internet maps on p. 79 courtesy of Delaware County, Ohio's, DALIS Project.

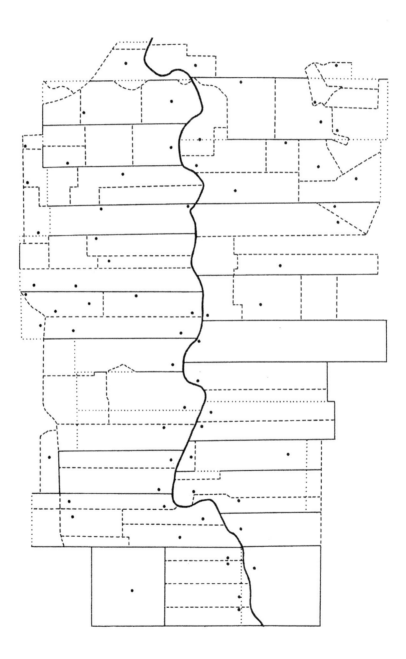

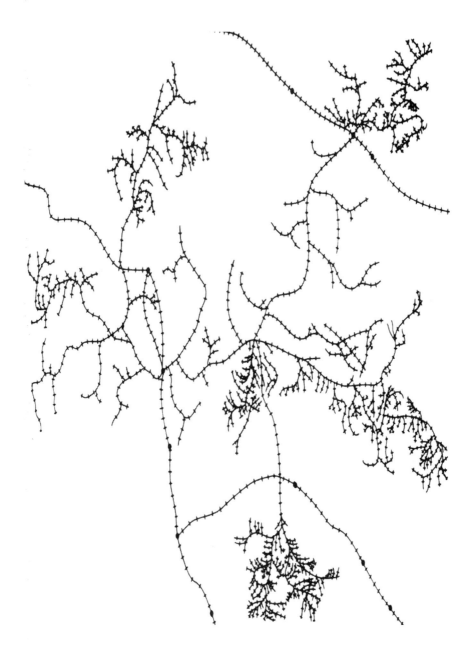

*How do you flatten,*
*shrink, and locate data?*

# Geographic Framework

***We have to mess with our spherical earth to get it flat.***
Through a process called "map projection" the curved surface of the
earth is flattened. We are used to seeing the flat earth on maps, but
what if we flattened a human body? Artists Lilla Locurto – 3D to the
left and 2D above – and Bill Outcault scanned the entire surface of their
bodies with a 3D object scanner. They brought the data into GeoCart
map projection software, and projected it. It is bizarre to see the human
body projected instead of the earth. We have clearly become accustomed
to the radical transformation involved in projecting the earth.

# Geographic Framework

We use maps to flatten the curved surface of the earth, shrink it down to a size we can handle, and systematically locate things.

We make maps for particular reasons, and those reasons guide the selection of a particular geographic framework - a map projection (which gets us flat), an appropriate scale, and a coordinate system (which helps us locate things on the map).

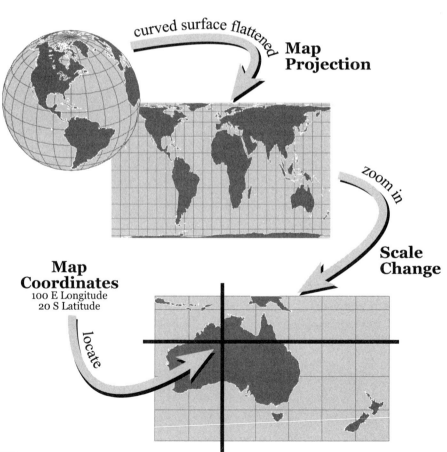

curved surface flattened

**Map Projection**

zoom in

**Scale Change**

**Map Coordinates**
100 E Longitude
20 S Latitude

locate

Choosing a **geographic framework** involves:

**Map Projection:** Flattening the earth's curved surface, and the resulting distortions of area, shape, distance, or direction

**Map Scale:** How much of the earth to show, and at what size?

**Map Coordinates:** Graph-like grids placed on the earth to assist with locating phenomena

x, y        91° W, 61°N

# ① Map Projection

Our earth's surface is curved. Most maps are flat. Transforming the curved surface to a flat surface is called map projection. All projected maps are flat, compact, portable, useful, and always distorted. Any curved surface gets distorted when you flatten it.

An orange peel *tears* when you peel and flatten it.

A toad skin *tears* when you peel and flatten it.

*Strike flat the thick rotundity o' th' world!*
William Shakespeare, *King Lear*

The surface of the Earth *tears* when you peel and flatten it. Peel a globe and you will get globe gores (below).

Most map projections stretch and distort the earth to "fill in" the tears. The Mercator projection (bottom) preserves angles, and so shapes in limited areas, but it greatly distorts sizes. Look at the size of Greenland on the globe compared to the Mercator.

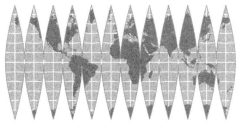

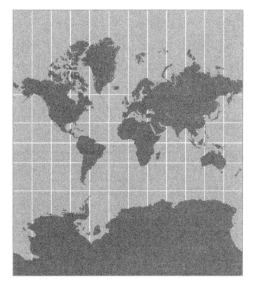

# distorting circles

*In the 19th century, Nicolas Auguste Tissot developed his "indicatrix," which can be used to evaluate map projection distortion. Imagine perfect circles of the same size placed at regular intervals on the curved surface of the earth. These circles are then projected along with the earth's surface. Distortions in the size and shape (angular distortion) of the circles show the location and quality of distortions on the projected map.*

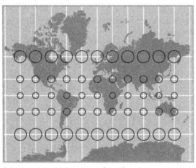

**Left:** Tissot's circles change *size* as you move north and south of the equator on the Mercator map projection. The more distorted the circles, the more distorted the areas of the land masses. Circle *shapes* are not distorted.

**Below:** Tissot's circles change *shape* over the surface of this area-preserving map. The more distorted the circles, the more distorted the shapes of the land masses. Circle *sizes* are not distorted.

**Mercator Map Projection:**
Preserves shapes, distorts areas.

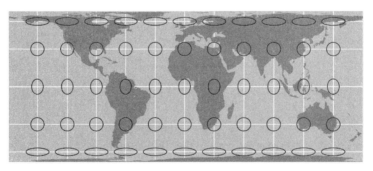

**Equal-Area Map Projection:** Preserves areas, distorts shapes.

96

# distorting bodies

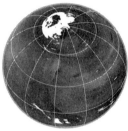

We can use Bill Outcault's projected body to more clearly see map projection distortions. Whenever you look at a projected map of the earth, think about what is happening to the earth. Cripes!

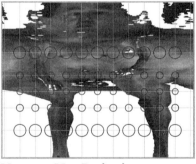

**Left:** Tissot's circles change size as you move north and south of Bill's waist on the Mercator map projection. The more distorted the circles, the more distorted the areas of the body.

**Below:** Tissot's circles change shape over the surface of this area-preserving map. The more distorted the circles, the more distorted the shapes of different parts of the body.

**Mercator Map Projection:**
Preserves shapes, distorts areas.

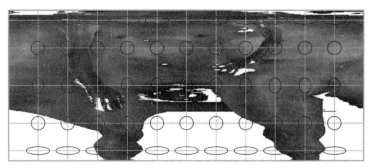

**Equal-Area Map Projection:** Preserves areas, distorts shapes.

# distorting data

*Mappable data is always associated with a location on the earth's surface. That is, mappable data is always tied to the grid. Because this grid gets distorted when it's projected from the curved surface of the earth to the flat surface of the map, the data tied to the grid gets distorted too.* **Projections matter because of what they do to our data!** *Since map projections do things to our data, it's important that what they do to the data clarifies, not muddies, it.*

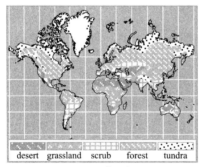

desert  grassland  scrub  forest  tundra

**Mercator Map Projection:**
Preserves shapes, distorts areas.

**Left:** A map of vegetation on a **Mercator** distorts the data. Northern vegetation types are greatly expanded in area compared to vegetation types near and south of the equator. This visually suggests the global dominance of northern vegetation types, which is not correct.

**Below:** The same vegetation data on a map projection that **does not distort areas.** But now shape is distorted! You have to be smart with projections and understand tradeoffs. In this case, with area vegetation data, you are better off distorting shapes and not areas.

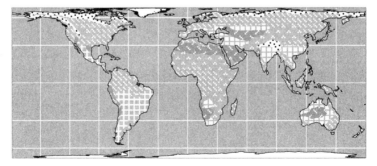

**Equal-Area Map Projection:** Preserves areas, distorts shapes.

# distorting layers

*We often impose multiple layers of mappable data on top of one another.  Since all mappable data are tied to a grid, we have to have the same projection for each layer of data for all of the layers to align properly.  This is important when combining digital layers in GIS, but also when compiling data from existing map sources.  Always try to learn the projection of your data sources.*

Two map layers, each with a *different* projection: an **Albers equal area in light grey** and a **Mercator conformal in dark grey.**  The data will not properly align!  This causes visual and computational problems.

Use GIS tools to *transform* one of the layers so both layers have the same projection.

# what projections preserve

*No map projection preserves the attributes of a globe, which maintains area, shape, distance, and direction. Map projections can preserve one or two of the attributes of the globe, but not all. Select a map projection that makes the best sense for your data.*

## preserving area

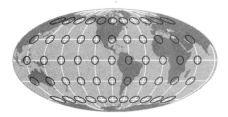

Some projections preserve size. This means that areas which are the same size on the globe are the same size on the map.

Area preserving, or equal-area map projections, are a good default for most maps.

**Mollweide Projection**
Oval shape, preserves area. Rounded map shape suggests the roundness of the earth. Projection can be **recentered** to minimize shape distortions of regions of greatest interest.

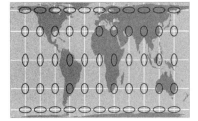

**Gall Projection:** Also known as the Peters Projection. This area-preserving projection's straight grid lines make north-south relationships straightforward. Rectangular shape makes it easy to fit into page layouts. Excellent projection for illustrating the shape distortions inherent in area-preserving map projections, and the area distortions inherent in shape-preserving projections.

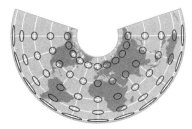

### Albers Equal-Area Projection

A common area-preserving map projection. **Poor** for world scale maps because of shape distortion and peculiar form. However, r**ecentering** on the area of interest (the U.S.) and selecting **part of the earth** (continent, country) results in an equal-area map with minimal shape distortion.

**Good** for regional maps, particularly when mapping area data.

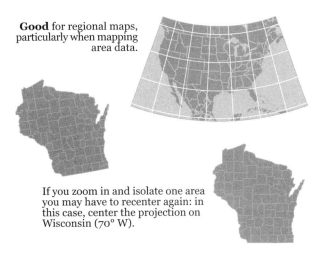

If you zoom in and isolate one area you may have to recenter again: in this case, center the projection on Wisconsin (70° W).

## *preserving shape*

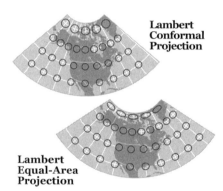

**Lambert Conformal Projection**

**Lambert Equal-Area Projection**

At sub-global scales distortions of area and shape are not as visually evident.

Saying that certain map projections preserve shape is not technically correct, but makes sense to normal people. As long as you are away from areas of high distortion, the shapes of contintents look OK. Conformal map projections preserve angles (around points) and therefore shape in small areas. Sizes are consequently distorted.

Conformal map projections are good for mapping regions and continents, especially when statistical data are involved.

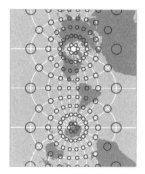

**Transverse Mercator**

On the Mercator projection, scale is true along the equator. When that projection is recentered sideways along a meridian - or line of longitude - scale is true along that meridian. This recentered projection is known as the Transverse Mercator and is the basis for the Universal Transverse Mercator coordinate system. When areas **a few square miles in size** are mapped using this projection, they are effectively free of all distortion.

## Mercator Projection

The Mercator is one of the few conformal world projections. Its distortions of sizes are nasty and it is a **poor** choice for a world map.

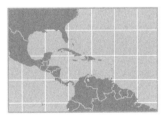

**Good** for regional sailing and flying charts, as any straight line drawn on a Mercator is a true compass bearing.

**Good** for maps of very small areas. Many detailed topographic maps are based on the (transverse) Mercator.

# *preserving distance / direction*

Stretch a piece of string between two points on a globe, and you will get the shortest distance between the points (called a great circle). Some projections preserve such distance relations: a straight line between two points on the map is the shortest distance between those two points on the earth. Distance relations cannot be preserved on equa-area maps.

Direction can be preserved on area-, shape-, or distance-preserving map projections.

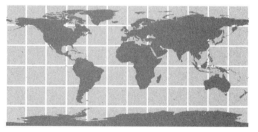

**Geographic Coordinate System, aka the "Geographic Projection."** The default projection in a popular GIS software package, similar to the Plate Carrée projection. This unpleasant item preserves nothing but distance, principally for north-south measurements. This projection is increasingly common, especially on internet mapping sites, where it distorts the data mapped on it.

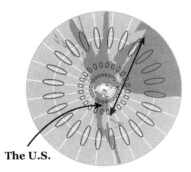

**The U.S.**

**Gnomonic Projection:** A straight line drawn anywhere on a Gnomonic projection is a great circle route, the shortest distance between two points. Terrifying distortions of area and shape and the inability to show more than half the earth at a time limit other uses of this projection.

### Azimuthal Equidistant Projection

Planar (azimuthal) map projections preserve directions (azimuths) from their center to all other points. The azimuthal equidistant projection also preserves *distance*: a straight line from the projection center to any other point represents accurate distance, in addition to correct direction and the shortest route. **Great** for offices of travel agencies when centered on their hometown.

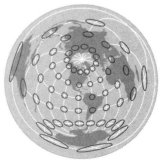

**Poor** for showing anything but distances from a point. Areas and shapes are wildly exaggerated as you move from the center.

105

# *interruptions*

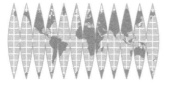

Globe gores, peeled from a globe and flattened, are akin to interrupted map projections. Interrupted map projections minimize distortions on the uninterrupted part of the map, and are typically used on maps of the entire earth.

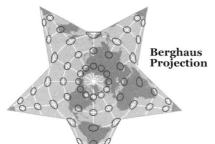

**Berghaus Projection**

Interrupted map projections are commonly used for maps of global statistical data. They have also been used as cute icons (Berghaus Star) and as the basis of "cut and assemble" globes (Fuller). The Berghaus is equidistant north of the equator. The Fuller has constant scale along the edges of all twenty of the triangular pieces. Within each of these triangles, size and shape are well presented.

**Fuller Projection**

### Goode's Homolosine Projection

Goode's is a common interrupted map projection used for world maps of statistical data. The projection does not distort areas, and shape distortions in the uninterrupted areas of the map are minimized. *Interruptions can be moved:* a map for ocean phenomena can interrupt land areas:

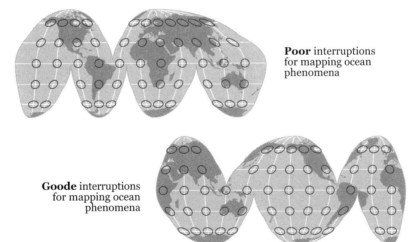

**Poor** interruptions for mapping ocean phenomena

**Goode** interruptions for mapping ocean phenomena

## *most of everything*

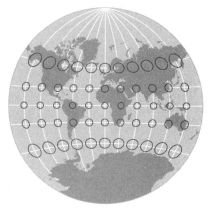

Map projection is most visible at a global scale, where distortions of areas and shape are most evident. Area-preserving projections often badly distort shapes, and shape-preserving projections area. But there is an alternative.

**Compromise** map projections distort area and shape a bit, but neither too badly. Compromise is *a good thing.*

The **Van der Grinten Projection** does not preserve shape or area, but minimizes their distortions in all but polar regions. Usually the polar regions are lopped off and the map presented as rectangular.

### Robinson Compromise Projection

Arthur Robinson's map projection preserves neither area nor shape, but reduces the distortion of both. The projection has rounded sides, suggesting the spherical earth, and avoids excessive distortion near the poles.

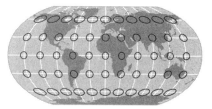

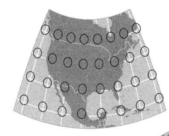

### Winkel Tripel Compromise Projection

This projection resembles the Robinson, but it has less areal exaggeration in the polar regions. The Winkel Tripel is the default world projection for the National Geographic Society.

**Poor** for regional or local scale maps because of area and shape distortion.

**Good** for a general world map and for mapping global phenomena

## 2 Map Scale

The earth is big.  Maps are small.  Map scale describes the difference, verbally, visually, or with numbers. Map scale will be determined by your goals for your map.  Map scale affects how much of the earth, and how much detail, can be shown on a map.

*verbal*

*visual*

*small scale*

*large scale*

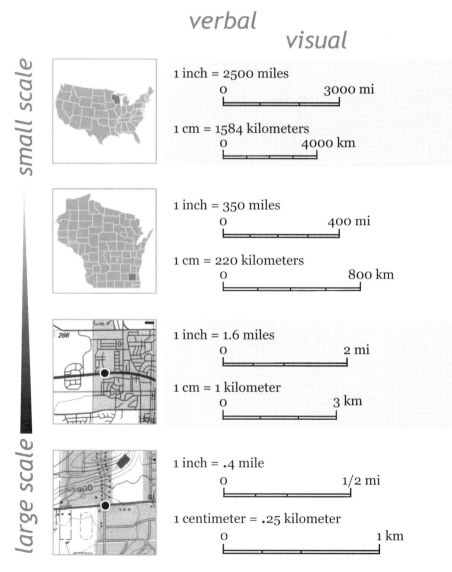

1 inch = 2500 miles

0                  3000 mi

1 cm = 1584 kilometers

0              4000 km

1 inch = 350 miles

0                  400 mi

1 cm = 220 kilometers

0                  800 km

1 inch = 1.6 miles

0                  2 mi

1 cm = 1 kilometer

0                  3 km

1 inch = .4 mile

0                  1/2 mi

1 centimeter = .25 kilometer

0                  1 km

## numerical

### 1 : 155,000,000

A representative fraction (RF) shows the proportion between map distance and earth distance for *any unit of measure*. **1 inch** on the map is **155 million inches** on the earth. **1 cm** on the map is **155 million** cm on the earth.

### 1 : 22,000,000

Distortions from map projections become less visually noticeable at regional and local scales. These distortions may become evident when combining map layers with different projections in GIS.

### 1 : 100,000

*Divide* a representative fraction:

    1 / 100,000 = .00001
    1 / 24,000   = .00004

The former is *smaller* than the latter: thus 1:100,000 is *smaller* scale than 1:24,000 (*larger* scale).

### 1 : 24,000

Larger-scale maps show *more detail* but of a *limited area*. Map projection distortions are less evident and distance is relatively accurate over the entire map.

*small scale*

*large scale*

# Map Coordinates

Map coordinates are a pair of numbers or letters which locate a point on a map. Linear features, such as rivers, are located as a string of connected point coordinates. Area features, such as a country, are located as a closed string of connected point coordinates. There are many different map coordinate systems.

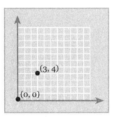

Map coordinate systems are based on the idea of a grid. A point is given a *location* on the grid in relation to an origin (0, 0). Lines and areas are defined by connecting a series of points.

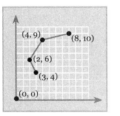 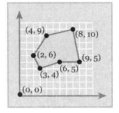

## Where is (0, 0)?

Where, on earth, should the origin (0, 0) be? If Washington, DC, is the origin, then all other locations are in relation to Washington. *Different coordinate systems have different origins.*

## Flat or sphere?

Coordinate systems which cover part of the earth assume a flat earth to take advantage of the easier *planar geometry*. Coordinate systems which cover all of the earth assume *spherical geometry*.

## Area covered?

How much of the earth is covered by the coordinate system? All of it? Part of it? *Different coordinate systems cover all or only part of the earth.*

## Units?

Coordinate systems can be in English units (feet), metric units (meters), or degrees. *Different coordinate systems have different units of measurement.*

112

# matching coordinates

*Map readers may need to understand common map coordinate systems to locate features on maps. Map makers need to understand map coordinate systems. When two or more digital maps are combined (as layers in GIS) they must have the same coordinate system, or they will not align properly. When new data are created (GPS or digitizing) they must be in the same coordinate system as other digital maps they are to be used with.*

**Poor coordinate system choice:**

**Delaware County, OH:** A digital map of county roads is in the State Plane Coordinate System (SPCS). A second digital map shows county buildings on the National Historical Register. It uses UTM (Universal Transverse Mercator) coordinates. Combining these maps in GIS without a common coordinate system results in misaligned map layers.

**Good coordinate system choice:**

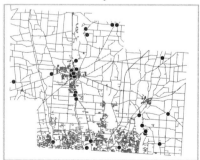

Transforming the Historical Register layer (using GIS tools) into SPCS allows the two maps to be properly aligned. SPCS is chosen in this case because many other Delaware County digital maps use SPCS (a common choice for regional planners in the U.S.).

# latitude & longitude

*Latitude and longitude cover the entire earth with one system and a single origin. In this system, locations are specified in degrees, minutes, and seconds.*

**90° N**

**0°**

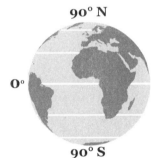

**90° S**

The equator is the origin for **latitude**. Lines of latitude are called **parallels**. Parallels run east-west, measuring 90° north and 90° south of the equator. Parallels never converge: one degree of latitude is always 69 miles (111 km).

**0°**

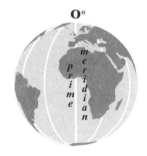

Greenwich, England, is the origin (prime meridian) for **longitude.** Lines of longitude are called **meridians.** Meridians run north-south, measuring 180° east and 180° west of the prime meridian. Meridians converge at the poles: one degree of longitude at the equator is 69 miles (111 km); one degree of longitude at the poles is 0 miles/km (a point).

**(0,0)**

The single origin (0, 0) is off the coast of Africa. Coordinates fall into one of four quadrants to the N/S (latitude) and E/W (longitude) of this origin.

Latitude/Longitude is measured in *degrees:* there are 60 minutes in 1 degree and 60 seconds in 1 minute.

Latitude/Longitude is used when you need a single coordinate system for the entire earth.

# universal transverse mercator

*The Universal Transverse Mercator (UTM) coordinate system is associated with the U.S. military. UTM, based on the "transverse" (sideways) Mercator projection, covers the entire earth, which is divided into 60 zones, each 6° wide, running from pole to pole. Planar geometry (a flat earth) is assumed.*

**Zone 17**

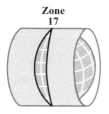

UTM zones are widest at the equator and narrow to a point at the poles. Ten zones (10 - 19) cover the continental U.S.

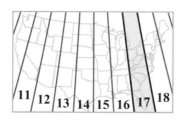

11 12 13 14 15 16 17 18

Each 6°-wide zone has a North and South zone. An origin (0, 0) is established to the south-west of each zone, so that all coordinates in the zone are positive.

UTM is measured in *meters*. A point is located in terms of how many meters east and north it is from the origin.

UTM is used by environmental scientists, the military, and any other professions who work at a regional or local scale but need their maps to coordinate with maps of other areas on the earth.

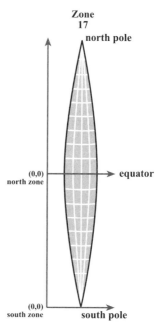

**Zone 17**

north pole

(0,0) north zone — equator

(0,0) south zone — south pole

# *state plane coordinates*

*The State Plane Coordinate System (SPCS)
was developed by the U.S. Coast and Geodetic
Survey. SPCS is used only in the United States,
which is divided into several hundred areas,
each with its own coordinate system. Since
each area is relatively small, planar geometry
(a flat earth) is assumed. Similar coordinate
systems are used in other parts of the world.*

Small U.S. States have a single **SPCS**
zone; larger states are divided into
several zones. No zone crosses a
state boundary.

Wisconsin has three SPCS zones.
An origin (0, 0) is established to the
south-west of each zone, so that all
coordinates in the zone are positive.

SPCS is measured in *feet*. A point
is located in terms of how many feet
east and north it is from the origin.

SPCS is used by planners, urban
utilities, environmental engineers,
and other professions who work at
a regional scale.

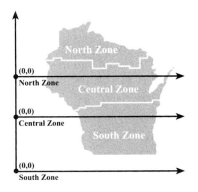

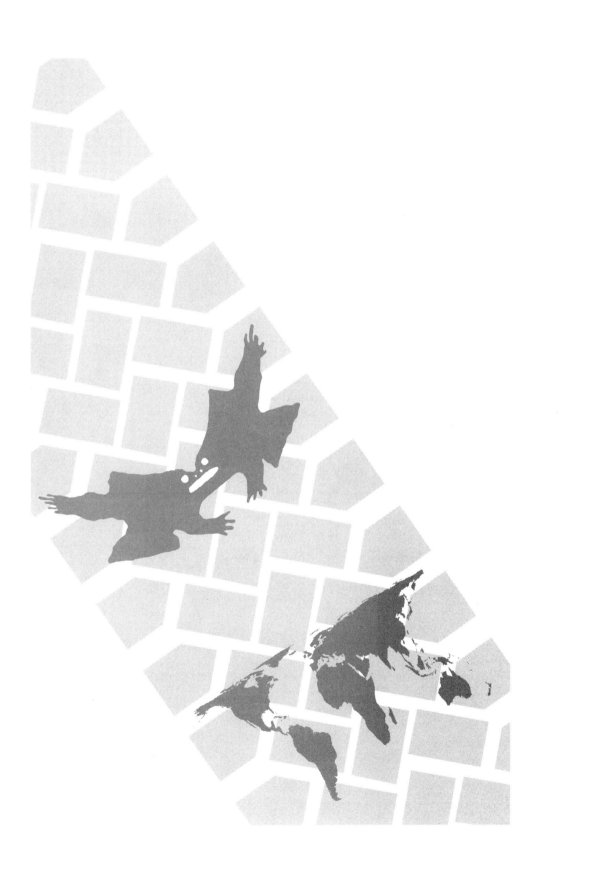

An unseen ruler defines with geometry
An unrulable expanse of geography
An aerial photographer over-exposed
To the cartologist's 2D images knows
The areas where the water flowed
So petrified, the landscape grows
Straining eyes try to understand
The works, incessantly in hand
The carving and paring of the land
The quarter square, the graph divides
Beneath the rule a country hides

Interrupting my train of thought
Lines of longitude and latitude
Define and refine my altitude

Wire. *Map Ref. 41°N 93°W.* (song lyrics, 1979)

"According to the map we've only gone 4 inches."

Harry Dunne. *Dumb & Dumber.* (movie, 1994)

"Just a map! Maps, my dear, are the undergarments of a country!"

Morgan the Goat. *The Englishman Who Went Up a Hill But Came Down a Mountain.* (movie, 1995)

# more information...

An accessible introduction to map projections is *Seeing Through Maps: The Power of Images to Shape Our World View,* by Ward Kaiser and Denis Wood (ODT, 2001).

A great handbook describing and illustrating dozens of map projections is *An Album of Map Projections,* published by John Snyder and Philip Voxland as a U.S. Geological Survey Professional Paper (#1453, 1989).

The best book on map projections is John Snyder's *Flattening the Earth* (University of Chicago Press, 1997). Snyder's approach is historical and exhaustive. The treatment is technical but not intimidating.

Every cartography textbook has a chapter or two on map projections and coordinate systems. The map projection chapter in Borden Dent's *Cartography: Thematic Map Design* (Brown, 1998) has a good chapter on map projections, and Phillip and Juliana Muehrckes' *Map Use* (JP Publications, 1998) has a good chapter on coordinate systems.

**Sources:** The body projection on pp. 88-91 and 97 courtesy of Lilla Locurto and Bill Outcault. The majority of map projections in this chapter were generated in GeoCart software, and a few in ArcGIS. The idea for the flat toad (p. 94) is borrowed from Edward Tufte, *The Visual Display of Quantitative Information* (Graphics Press, 1983). The map of vegetation on a Mercator map projection (p. 98) is based on a map in Anne Spirn's *The Granite Garden: Urban Nature and Human Design* (Basic Books, 1984). Delaware County, Ohio, data (p. 113) courtesy of the DALIS Project. The State Plane Coordinate and Universal Transverse Mercator diagrams (pp. 115-116) were re-created from the Muehrckes' *Map Use* (1998) and Kraak and Ormeling's *Cartography* (Pearson, 1996). Lyrics to *Map Ref. 41°N 93°W* written by Graham Lewis, Colin Newman, Bruce Gilbert used by permission of Carbert Music Inc.

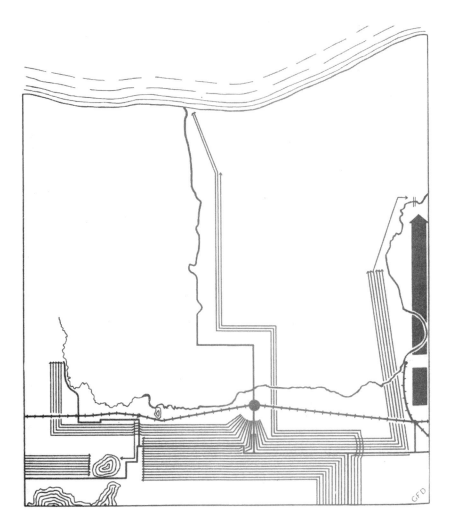

# Queer Geography of Columbus, Ohio

Columbus has the largest number of gay and lesbian couples in the state. The city has a reputation as "gay-friendly" and tolerant. Gay residents of Columbus play an important role in business and the arts, and have settled in particular neighborhoods. The *queer geography* of Columbus is distinct: while lesbians and gay men are found everywhere in the region, there are distinct lesbian and gay neighborhoods. Lesbian neighborhoods tend to have less expensive housing, are more residential, and are in safer areas away from downtown. Gay neighborhoods have more expensive housing, are near entertainment and other urban amenities, and are located near downtown. Distinct *spatio-sexual differences* reflect generalized characteristics of the gay and lesbian community: lesbians are domestic and have lower incomes; gays are less domestic and have more expendable income.

"Lesbians drift north."
Gwen

Far **north Clintonville** is home to younger lesbians, who find the 1950s Cape Cod homes affordable, and the neighborhoods quiet and safe.

*old suburban $$*
*not gentrified*

**Clintonville** is home to many lesbian families, who find it a safe and affordable place to live and raise families. Clintonville is an old Columbus suburb, developed in the 1920s, and dominated by moderate-sized homes. Clintonville is not diverse racially, but is home to blue-collar and white-collar families, students (from Ohio State University, to the south), and retirees.

*old suburban $$*
*not gentrified*

**Ohio State University**

**The University District,** adjacent to Ohio State University, is home to younger lesbians who rent or own homes in the student-dominated area. The neighborhood is less safe but affordable. The lone lesbian bar outside of downtown is in this neighborhood.

*old suburban $*
*not gentrified*

**Victorian Village** is home to numerous gay singles and families. Located between Ohio State and downtown, the area's grand Victorian homes were converted into apartments in the mid-20th century, prior to gentrification beginning in the 1980s. The area is known for its restaurants, clubs, bars, and art galleries - just to the east, along High Street, in the Short North.

*urban $$$$*
*gentrified*

**Italian Village** is an area of early 20th century homes to the north-east of downtown. It has a significant number of African Americans, but is being gentrified in part by gay singles and couples. The neighborhood is close to the entertainment amenities of the Short North, just to the west.

*urban $$*
*gentrifying*

**Down-town**

**Westgate** is a "new" area for gay and lesbians. Located in the largely working-class west side of Columbus, it's affordable 1920s homes make it an affordable alternative to

*old suburban $*
*gentrifying*

**German Village** is home to numerous gay singles and families. The area is a gentrified neighborhood of late 19th century homes, tightly packed together in a very urban setting just south of downtown. Excellent restaurants, clubs, and bars as well as very expensive housing distinguish this neighborhood.

*urban $$$$*
*gentrified*

**Olde Town East** is home to gay singles and families who are gentrifying the inexpensive Victorian housing stock. The area just east of downtown, is predominantly African American and, as with many gentrifying areas, tensions are evident. The homes are grand, as this area was once the home of prominent Columbus families.

*urban $$*
*gentrifying*

L = Lesbian  G = Gay Men     Size of letter = proportion of G vs. L      $ = cost of housing

*Scale:* about 6 miles wide and 10 miles high        *Source:* Various gay & lesbian folks in Columbus, U.S. Census 2000 Data

*How is it organized?*

# Columbus, Ohio
# Same Gender, Living Together, 1999

Total
Persons

 93-146 | 66-93 | 44-66 | 27-44 | 1-27 | No Data

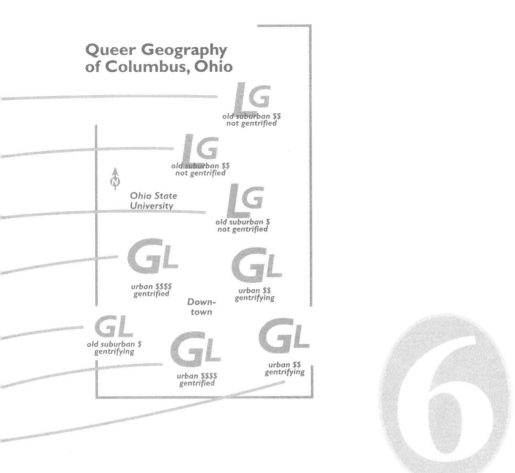

**Queer Geography of Columbus, Ohio**

old suburban $$
not gentrified

old suburban $$
not gentrified

Ohio State University

old suburban $
not gentrified

urban $$$$
gentrified

urban $$
gentrifying

Down-town

old suburban $
gentrifying

urban $$$$
gentrified

urban $$
gentrifying

# Map Layout

***There is more to a map than the map.*** Maps often have titles, a legend, explanatory text, scale and directional indicators, sources, credits, a border, insets, and locator maps. Arranging these "map pieces" around and upon the map is map layout. Effective use of map pieces and layout enhances the goals of your map. A "map" of the queer geography of Columbus, Ohio, *doesn't even include a map,* but the *map pieces* (particularly the explanatory text) and *layout* (neighborhood descriptions are located approximately where they are in Columbus) impart a large amount of information about the urban gayscape. The 1999 data on people of the same gender living together (left) from the U.S. Census reveals similar patterns in a different form.

# Map Layout

Maps typically include a title, legend, scale, explanatory text, directional indicator, sources and credits, a border, insets, and locator maps. These *map pieces* are systematically arranged around and upon the map.

Consider all the elements you will include on your map, as well as the overall map layout, early in the map-making process. Link decisions on map layout to your goals for your map and the final medium. Together these will shape your choices of map content, scale, and the kind of information you include on and surrounding your map.

Map layout is intuitive and some people are better at it than others. It is relatively easy to experiment and create good map layout, particularly with map-making software. When map layout *succeeds,* map readers will *not* notice: they will focus on the content of the map. When map layout *fails,* the map reader *will* notice: an awkward layout distracts the map reader from the subject and goals of the map.

**Poor layout:**

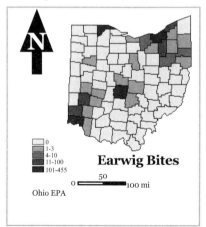

**Good layout:**

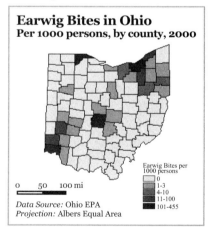

**Map layout** consists of:

**Map pieces:** title, legend, scale, explanatory text, directional indicator, sources, credits, border, insets, and locator maps.

**Focus:** How do map readers' eyes scan across a map, and where do they typically focus?

**Balance:** How can map elements be balanced to enhance understanding of the map?

**The grid:** How can a layout be proportioned into horizontal and vertical spaces to enhance understanding of the map?

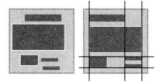

# **1** Map pieces

A series of map elements which most maps include. Map layout consists of arranging these "map pieces." Which map pieces you include depends on your goals for the map.

## *title*

Map titles vary greatly, but should attempt to include:

- ✓ what: the *topic* of the map.
- ✓ where: the *geographic* area.
- ✓ when: *temporal* information.

Titles are important map elements, and type *size* should be *two to three times* the size of the type on the map itself. A subtitle, in smaller type, is appropriate for longer titles or more complex map subjects.

**Poor title content:**

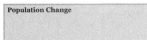

**Poor title size:**

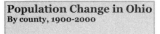

**Good title content and size:**

**Population Change in Ohio**
By county, 1900-2000

## *legend*

Map legends vary greatly, but should include any map symbol you think may not be familiar to your audience.

The legend is the *key* to interpreting the map. If it fails, your goals for the map will fail. However, don't insult your map's readers by including obvious symbols in the legend. You need *not* preface the legend with "Legend" or "Key" as most map readers know that without being told.

## *scale*

Most large- and medium-scale maps (of a limited area up to continental scale) should include a scale indicator, particularly if your map's readers need to make measurements on the map.

Verbal and visual scales are more intuitive, numerical scales more flexible (they can be metric or English). If your map's users might reduce or increase the size of the map, the visual scale is the best option (it will remain accurate even if scaled).

mi
1 inch = 1000 mi
**1 : 1,200,000**

Small-scale maps (of the entire earth or a substantial portion of it) should not include a simple visual scale because such maps always contain substantial scale variations.

# explanatory text

Explanatory text is often vital to the success of your map.

You cannot express everything you need your map's readers to understand with the map itself. Use text blocks on the map to communicate information about the map content, its broader context, and your goals.

- ✓ on a historical map, include a paragraph setting the historical context, and also include blocks of text on the map that tell what happened at important locations.

Explain your interpretation of your map's patterns with text: tell your map's readers (in addition to showing them with the map) what you think about the mapped data.

- ✓ on a map showing changes in average income over the last decade (by county in a state), explain your interpretation that suburban counties are becoming richer, and urban and rural counties poorer, due to recent tax cuts.

The readers of your map may agree or disagree with your interpretation, but your interpretation and intent will be clearly communicated.

# directional indicator

A directional indicator is needed on a map in two cases:

- ✓ the map is *not* oriented north.
- ✓ the map is of an area *unfamiliar*. to your intended audience.

Directional indicators can often be left off the map. If included, avoid large and complex directional indicators: they are relatively unimportant map elements, and should not be visually prominent.

# sources, credits

Each map you make should include:

- ✓ data source(s).
- ✓ map maker, and when made.
- ✓ map projection and coordinate system information, particularly if you think someone may use your map in GIS (so they can coordinate your map with other GIS map layers).

# border

A border drawn around the map, title, legend scale, inset, and directional indicator may draw together your map. Or not - you may not need one.

If used, make the border narrow, possibly in grey: it will be noticeable but not too noticeable.

# insets and locator maps

Choosing a single map scale is difficult: at a smaller scale, data in one area on the map are too dense. At a larger scale, the dense data are understandable, but the geographic extent is limited - lopping off part of a state or country. In addition, readers of very detailed large-scale maps need a sense of where the area on the map is. Map insets and locator maps are a common way to deal with such problems by showing multiple scales simultaneously.

**Good use of inset:**

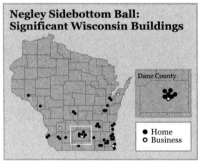

An inset which *jumps to a larger scale (less area, more detail)* helps the map reader to understand areas on the map where data are more dense and difficult to distinguish at the scale of the main map. Text or subtle arrows guide the map reader's attention from the main map to the inset. If the legend has different symbols from the main map, include an inset legend. It is also good practice to include a scale indicator for the inset.

**Better use of inset w/locator map:**

A locator map which *jumps to a smaller scale (more area, less detail)* helps the map reader to understand the geographic context of the area on the main map. A map of the U.S. or the earth, for example, helps map readers to see where Wisconsin is. U.S. map readers probably know the location of Wisconsin, while international readers may not. An *orthographic map projection* is commonly used as a smaller-scale locator map.

**Poor use of locator map:**

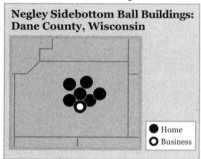

**Negley Sidebottom Ball Buildings: Dane County, Wisconsin**

Titles can be written to inform map readers of the broader geographic context of an area shown on a larger-scale map. However, map titles are often inadequate for conveying geographic location: where is Dane County in Wisconsin? If the map reader is unfamiliar with Wisconsin counties, the geographic context of this map is missing.

**Good use of locator map:**

A locator map which jumps to the next significant geographic area up (in scale) from the main map will help the map reader to contextualize the map. Larger-scale maps benefit greatly by including a smaller-scale locator map.

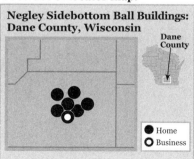

**Negley Sidebottom Ball Buildings: Dane County, Wisconsin**

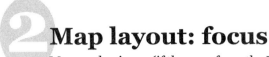

# Map layout: focus

Map readers' eyes (if they are from the Western world) typically follow a path over a map, eventually focusing near the center of the map. Use map layout to guide readers through map elements in the order you want, leading them to the most important part of the map.

## eye movement

*Reading a map is like reading a page in a book: you start in the upper left and end in the lower right. Most map readers scan the entire map in this manner prior to a more careful reading of the map. Position map elements so that those that should be seen first are in the upper left part of the map.*

**Poor focus:**          **Good focus:**

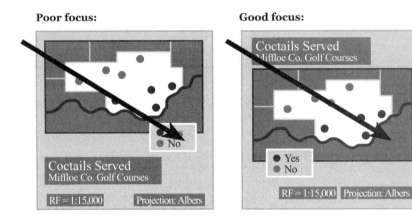

# visual center

*The visual center of a map layout is slightly above the actual center. Centering implies importance. Position map elements so that the most important are near the visual center of the map: the map reader will focus near this visual center, particularly if there is more detail, and assume the map elements there to be most important.*

**Poor centering:**

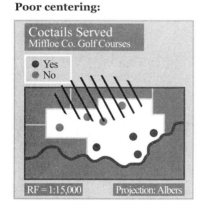

**Good centering:**

# Ⓐ Map layout: balance

Balance refers to the stability of a map layout. When balance is poor, map readers may be distracted. When balance is achieved, map readers will focus on the content of the map. Balance can be symmetrical or asymmetrical.

## *balance*

*Balancing map elements is complicated and intuitive. The map elements to balance vary in weight.* **Heavier** *elements include those that are larger, darker, brightly colored, simpler and more compact in shape, and closer to the map edge (particularly the top).* **Lighter** *elements include those that are smaller, lighter, dully colored, complex or irregularly shaped, and closer to the map center.*

**Poor balance:**

**Better balance:**

# *symmetry*

*Graphic designers trained in the U.S. tend to think about symmetry as having roots in Egyptian and Greek architectural ideals. Balance around a central vertical axis defines symmetrical layout in this way of thinking. Similarly they think about asymmetry as having roots in Oriental, especially Japanese, architectural ideals. Asymmetrical layout depends on off-center weights and balances.*

### Symmetrical Balance

### Asymmetrical Balance

### Symmetrical balance implies:

✓ tradition
✓ conservative look
✓ simplicity
✓ rule following

### Asymmetrical balance implies:

✓ modernity
✓ progressive look
✓ complexity
✓ creativity

# 4 Map layout: the grid

Map layout is enhanced by an underlying grid which proportions layout space horizontally and vertically. A grid establishes horizontal and vertical sight-lines which further enhance the stability of the layout. Without a grid, a balanced layout may seem disjointed.

## sight-lines

*Sight-lines are invisible horizontal or vertical lines which touch the top, bottom, or sides of map elements. Minimizing the number of sight-lines simplifies complexity, reduces disjointedness, and stabilizes and enhances map layout. This allows your map's readers to focus on the map's subject.*

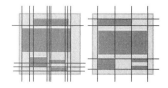

**Poor sight-lines:**

**Good sight-lines:**

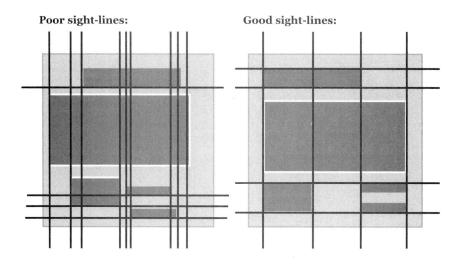

# grid symmetry

Grids establish sight-lines and divide space in such a way that focus and balance are enhanced. Grids for map layout can be symmetrical or asymmetrical. Symmetrical grids are based on two central axes (shifted slightly up to the visual center) and top, bottom, and side margins. Asymmetrical grids are more complex, but still depend on the visual center as well as maintaining top, bottom, and side margins.

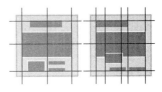

**Symmetrical grid:**

**Asymmetrical grid:**

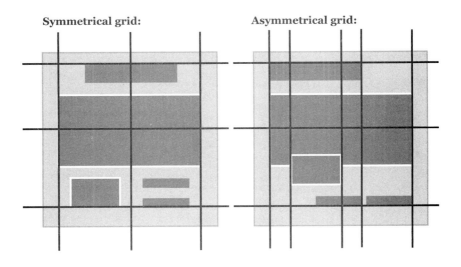

The superfluous irritates me sharply ... However, it is sometimes necessary to repeat what all know. All mapmakers should place the Mississippi in the same location, and avoid originality. It may be boring, but one has to know where he is. We cannot have the Mississippi flowing toward the Rockies for a change.

Saul Bellow. *Mr. Sammler's Planet.* (1970)

It had become a place of darkness. But there was in it one river especially, a mighty big river, that you could see on the map, resembling an immense snake uncoiled, with its head in the sea, its body at rest curving afar over a vast country, and its tail lost in the depths of the land. And as I looked on the map of it in a shop-window, it fascinated me as a snake would a bird – a silly little bird.

Joseph Conrad. *Heart of Darkness.* (1902)

You know how Venice looks upon the map –
Isles that the mainland hardly can let go?

Robert Browning. *Stafford.* (1837)

138

# more information...

As we move into map design, three wonderful books on the design of information graphics by Edward Tufte are must-looks: *The Visual Display of Quantitative Information* (Graphics Press, 1983), *Envisioning Information* (Graphics Press, 1990), and *Visual Explanations: Images and Quantities, Evidence and Narrative* (Graphics Press, 1997).

Paul Mijksenaar's *Visual Function: An Introduction to Information Design* (Princeton University Press, 1997) is a succinct and beautifully illustrated overview of information design – including map design.

Though it doesn't mention maps once, Philip Meggs' *A History of Graphic Design* (Van Nostrand Reinhold, 1992) is the standard history of European and American graphic design, with thousands of examples of excellent graphic layouts. Again, no maps, but Jaroslav Andel's *Avant-Garde Page Design* (Delano-Greenidge, 2002) illustrates one enviable layout after another.

*Cartographic Perspectives* is a journal that covers all aspects of mapping, and is available to members of the North American Cartographic Information Society (www.nacis.org).

**Sources:** The best source on many of the issues in this chapter is Borden Dent's *Cartography: Thematic Map Design* (1998). A few additional details are from Allen Hurlburt's *Layout: The Design of the Printed Page* (Watson Guptill, 1977).

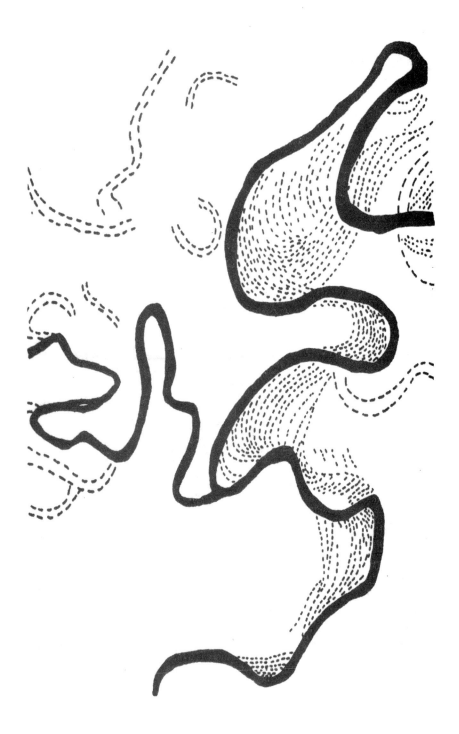

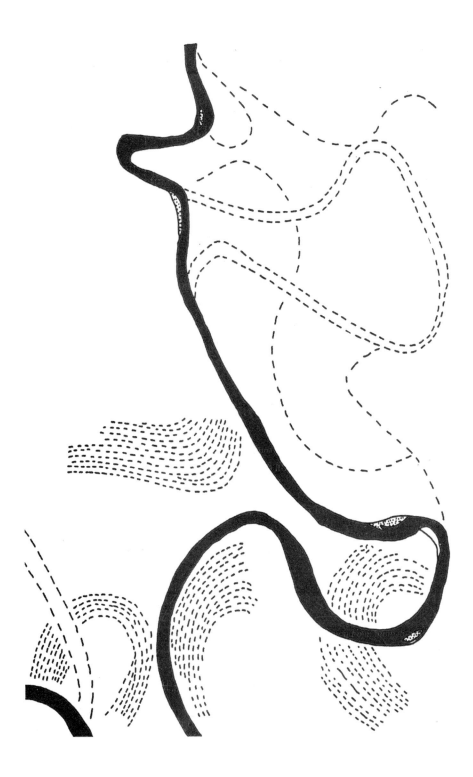

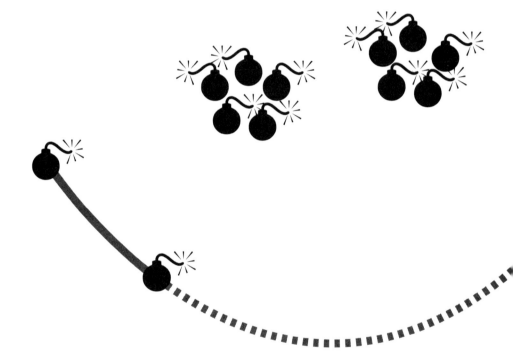

*What jumps out at you?*

# Geo-Smiley Terror Spree

Luke Helder, a university student from Minnesota, went on a two-week spree of bombings throughout the Midwestern U.S. in an attempt to create a giant smiley face across the nation.

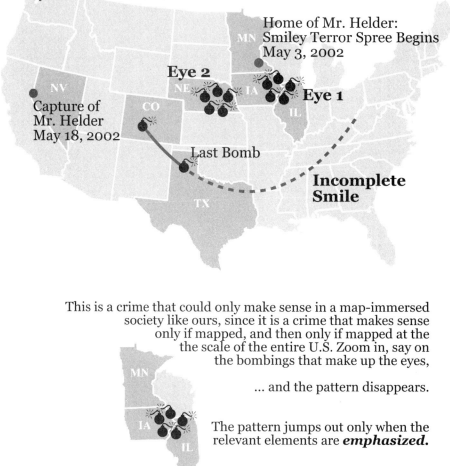

Home of Mr. Helder:
Smiley Terror Spree Begins
May 3, 2002

Eye 2

Eye 1

Capture of
Mr. Helder
May 18, 2002

Last Bomb

**Incomplete Smile**

This is a crime that could only make sense in a map-immersed society like ours, since it is a crime that makes sense only if mapped, and then only if mapped at the the scale of the entire U.S. Zoom in, say on the bombings that make up the eyes,

... and the pattern disappears.

The pattern jumps out only when the relevant elements are ***emphasized.***

# Intellectual and Visual Hierarchies

***Flat maps have depth.*** This map has many elements, but they are not all of equal importance. All maps have an *intellectual hierarchy* driven by your intent for the map. Some elements are vital to the goals of the map – telling the story of the geo-smiley bomber – and those elements are visually dominant. More important elements include bombing locations, the smile, and states involved in the spree. Less important elements, intellectually and visually, include states not involved in the spree and boundaries. Determine the intellectual hierarchy of elements on your map, then show it on the map with a distinct visual hierarchy.

# Intellectual and Visual Hierarchies

Your intent for your map drives its design. Intent suggests an **intellectual hierarchy**: what are the different elements on and around the map, and what are their relative importance? Once you have established a clear intellectual hierarchy, you can choose a **visual hierarchy** that reflects the intellectual hierarchy. If map elements are not important to your goals for your map, they are probably "map-crap" and can be left off.

## Depth on the flats...

Some elements **stand out,** and others **fall to the back.** This is **visual hierarchy.**

A successful visual hierarchy shows you what is most important **first**; these elements **jump out.** Less important elements are less visually noticeable and fall to the back. A successful visual hierarchy clearly communicates the intellectual hierarchy and **intent** of your your map.

Side view of graphic above showing depth.

**Poor visual hierarchy:**

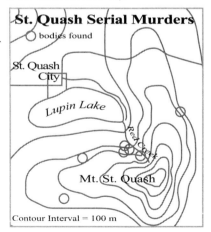

**Good visual hierarchy:**

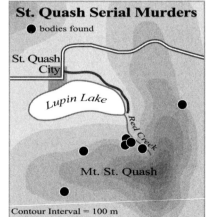

146

## Coordinating intellectual and visual hierarchies:

**1** **Figure-ground defined and illustrated:** What is figure-ground and how does it work?

**2** **Design guides for intellectual and visual hierarchies**

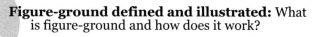

**3** **Enhancing visual hierarchies on maps:** How can depth be added to flat maps?

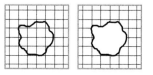

# ❶ Figure-ground defined and illustrated

The perceptual effect called figure-ground is behind our ability to see visual depth on maps.

Certain elements **stand out** in any image or graphic. These objects have **form** and are the **figure.**

*Figure-ground in a photograph*

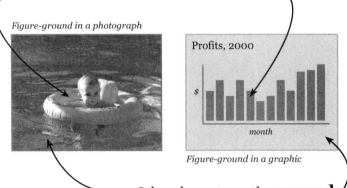

Profits, 2000

$ month

*Figure-ground in a graphic*

Other elements are the **ground.** These objects fall into the background and are less noticeable.

*Weak figure-ground in a map*

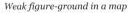
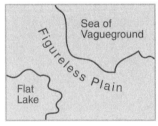

Sea of Vagueground

Figureless Plain

Flat Lake

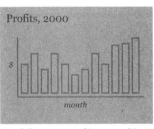

Profits, 2000

$ month

*Weak figure-ground in a graphic*

| **Figure on maps** | **Ground on maps** |
|---|---|
| ✓ most important map elements. | ✓ least important map elements. |
| ✓ more important meaning. | ✓ less important meaning. |
| ✓ distinct form and shape. | ✓ indistinct form and shape. |
| ✓ jump out. | ✓ fall back. |

Figure-ground is a powerful design element available to map makers. Manipulate figure-ground to drive home the point of your map.

Enhancing **figure-ground** and **visual hierarchy** is easy...

# Design guides for intellectual and visual hierarchies

Make a quick list of the elements on your map. Arrange them from *most* important in your map's intellectual hierarchy to *least* important. Select visual symbols (a visual hierarchy) which reflect these priorities.

A design guide (as below) reveals the ways you can visually manipulate point, line, and area symbols on a map to achieve visual depth.

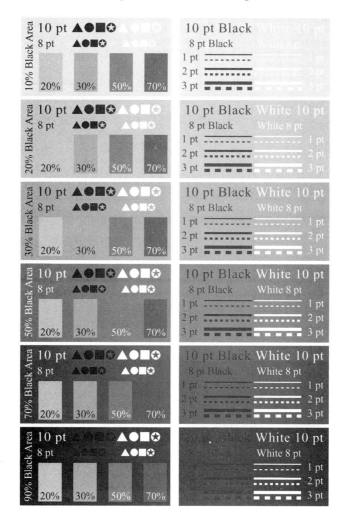

# ③ Enhancing visual hierarchies on maps

Add depth to flat maps to help the map reader
see the point of your map.

## visual difference

*Noticeable visual differences separate figure
from ground and enhance visual hierarchy.
The examples on the following pages all
enhance visual differences to build a visual
hierarchy. To focus attention on the most
important areas on your map, make it visually
different from peripheral areas.*

**Poor visual difference:**

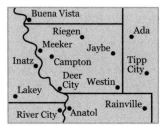

**Good visual difference:**

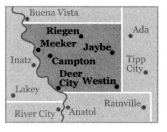

## detail

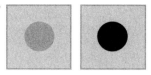

*Figure has more detail than ground. To focus
attention on the most important area on your
map, reduce detail in peripheral areas.*

**Poor detail:**

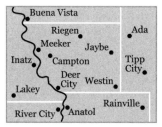

**Good detail:**

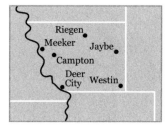

# edges

*Sharp, well-defined edges help separate figure from ground. Conversely, weakened edges move less important elements on the map from figure to ground. Grey or white (reversed-out) lines and type weaken edges and move less important information lower in the visual hierarchy.*

**Poor edges:**

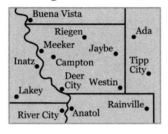

**Good edges:**

# layering

*Visual hierarchy is enhanced when the ground appears to continue behind the figure. Grids of latitude and longitude can often be visually manipulated to focus attention on the most important area of your map.*

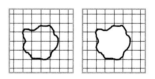

**Poor layering:**

**Good layering:**

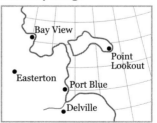

151

# texture

*Coarser textures tend to stand out as figure and move higher in the visual hierarchy. Map elements with the same orientation also tend to form a figure.*

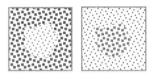

**Poor texture:**

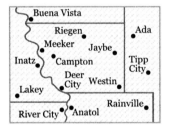

**Good texture:**

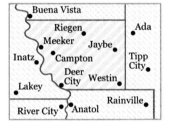

# shape & size

*Figure has shape and size. Map elements with simple closed shapes tend to be seen as figure. However, complex shapes also draw attention and tend toward figure. Larger symbols tend toward stronger figure.*

**Good shape:**

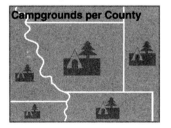

**Good shape:**

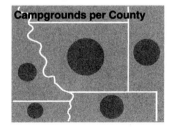

152

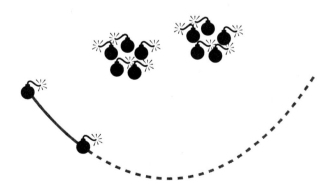

He loved maps, and in his hall there hung a large one of the Country Round with all his favorite walks marked on it in red ink.

J.R.R. Tolkien. *The Hobbit.* (1937)

Picture a map of India as big as a tennis court with two or three hedgehogs crawling over it....Each hedgehog might represent one of the dust storms which, during the summer, wander aimlessly here and there over the Indian plains, whirling countless tons of dust into the atmosphere as they go...until the monsoon rolls in and squashes them flat.

J.R. Farrell. *The Siege of Krishnapur.* (1973)

He looked on the map for the Lonely Tree, but did not find it.

"That map is absolutely worthless," said Good Fortune emphatically.

"Perhaps it is only a matter of looking long enough," Christian suggested. "Yesterday I found Durben Mot, and there's a lot of sand here.... We must be here, and the place is called Gatsen Mot, and there's a well near."

"That's wonderful!" cried Good Fortune, "and those who say a man should be sparing of his words are right. Gatsen Mot is Mongolian for Lonely Tree, and I beg your pardon for my hasty speech."

Fritz Muhlenweg. *Big Tiger and Christian.* (1954)

154

# *more information...*

One of the best discussions of intellectual hierarchy is Edward Tufte's in the second half of his book *The Visual Display of Quantitative Information* (1983). Another stimulating treatment is that of Richard Saul Wurman in his *Information Anxiety* (Doubleday, 1989). Wurman devotes a chapter to maps, but we recommend the book more generally. Borden Dent's *Cartography: Thematic Map Design* (1998) is an excellent overview of intellectual and visual hierarchies on maps.

A more general reflection on these issues in visual perception is provided by the essays in R.L. Gregory and E.H. Gombrich's *Illusion in Nature and Art* (Scribner's, 1973). In this collection, Jan Deregowski's essay offers a provocative cross-cultural perspective on the figure-ground problem in particular.

**Sources:** The "Geo-Smiley Terror Spree" map is based on data from a map in *Time magazine* (May 20, 2002). Much of this chapter is based on Dent's chapter on intellectual and visual hierarchies in his *Cartography: Thematic Map Design* (1998).

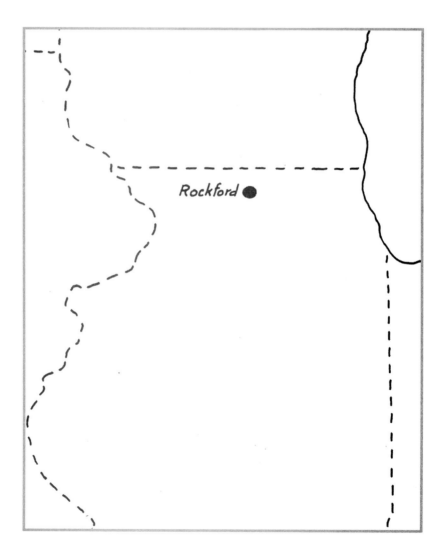

*Less is more?!*

*Arctic*

*Siberia*

*Gobi*

*Himalayas*

*Sahara*

*Indian*

# Map Generalization and Classification

***Fewer data are often better.*** Bill Bunge made a map he called *The Continents and Islands of Mankind* to make a point: as land and water barriers are about equally effective nowadays, there is no reason to keep mapping the continents and oceans when we are concerned with human affairs. This then is a map of places where there are more than 30 people per square mile in the world. The rest are oceans – land and sea. The shapes of the continents are obvious enough without being drawn. More importantly, they are beside the point. So leave them off the map! What matters is where people are. Fewer data more effectively make the point.

# Map Generalization and Classification

Our human and natural environments are complex and full of detail. Maps work by strategically reducing detail and grouping phenomena together. Driven by your intent, maps emphasize and enhance a few aspects of our world and deemphasize everything else.

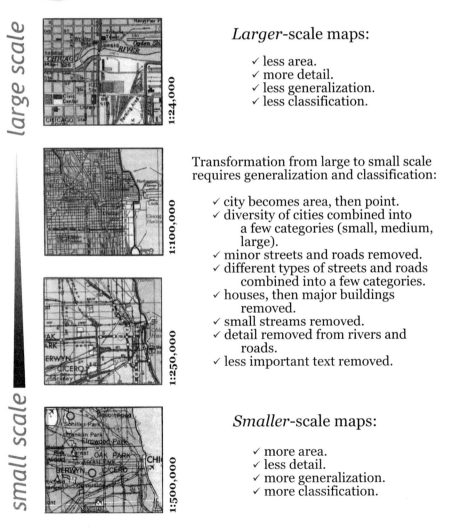

*large scale*

1:24,000

1:100,000

1:250,000

1:500,000

*small scale*

*Larger*-scale maps:

✓ less area.
✓ more detail.
✓ less generalization.
✓ less classification.

Transformation from large to small scale requires generalization and classification:

✓ city becomes area, then point.
✓ diversity of cities combined into a few categories (small, medium, large).
✓ minor streets and roads removed.
✓ different types of streets and roads combined into a few categories.
✓ houses, then major buildings removed.
✓ small streams removed.
✓ detail removed from rivers and roads.
✓ less important text removed.

*Smaller*-scale maps:

✓ more area.
✓ less detail.
✓ more generalization.
✓ more classification.

162

Two processes reduce detail on maps:

**Map Generalization:** the systematic reduction of detail to enhance the point of your map.

**Poor generalization:**

**Good generalization:**

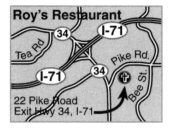

**Data Classification:** the systematic categorization of data to enhance the point of your map.

**Same data, *two* classifications ...**

**... two very *different* maps.**

# Map Generalization

More information on a map does not make it better!
Map generalization is driven by why you are making
your map.  Types of map generalization include
selection, simplification, smoothing, displacement,
enhancement, and dimension conversion.

## *selection*

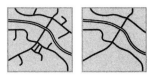

*Maps select a few (and don't select most)
features from the human or natural
environment.  Selected features are vital to
the intent of your map; unnecessary features
should not be selected.  Selection reduces
clutter and enhances the reader's ability to
focus on what is most important on a map.*

**Map Intent:** A small map for a phone book advertisement for Roy's
Restaurant.  The map must help people get to the restaurant.  Select only
the information that will help people get to Roy's: additional information
clutters the map and decreases clarity and communication.

**Poor selection:**                    **Good selection:**

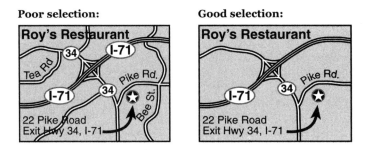

**Selection** is often the responsibility of the map maker.  Selection can be
automated with computer mapping.  For example, at a larger scale (more
detail, less area) all cities over 1000 in population are shown.  At a smaller
scale (less detail, more area) only cities over 100,000 population are shown.

### Questions to ask when making selections:

- ✓ is the feature necessary to make your point?
- ✓ will removing the feature make the map harder to understand?
- ✓ if less important features are removed, do more important
  features stand out more clearly?
- ✓ does removal of less important features lead to a less cluttered
  and busy-looking map?

# simplification

*Selected features are often simplified. Simplification can enhance visibility, reduce clutter, and, with digital data, reduce the size of the digital map file. Smaller-scale maps (showing more area) tend to have more simplified features than larger-scale maps.*

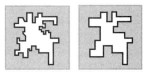

**Map Intent:** City boundaries can be quite complex. For an official city map, detailed boundaries are required, but for this map of Reedstown and surrounding towns for a business promotion, simplified boundaries are less distracting and better.

**Poor simplification:**   **Good simplification:**

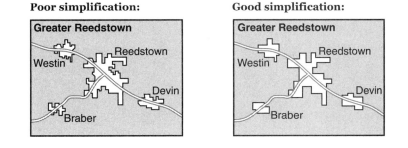

**Simplification** of features is done to the point where they are less complex, yet still recognizable. Identifying characteristics should be retained. Computers often simplify by methodically removing detail: a line is simplified by removing every other point, for example.

### Questions to ask when simplifying:

- ✓ how simplified can a feature be and still be recognized?
- ✓ does the removal of detail remove any vital information?
- ✓ does the simplification of a feature make it more noticeable?
- ✓ does the simplification of a feature make the map less cluttered and busy-looking?

# *smoothing*

*Smoothing map features reduces angularity. Smoothing involves simplification, but also adjustments in the location of a feature or possibly the addition of detail. Smoothing varies with the feature: rivers may be smoothed more than roads, as rivers tend to be smoother than roads.*

**Map Intent:** One of many small maps of parks and nature areas in a state guidebook for tourists. The data used for the river was poorly digitized, and does not capture the sinuous feel of a river. Smoothing entails removing some detail and adding points and character to the feature so that it looks more like a river.

**Poor smoothing:**

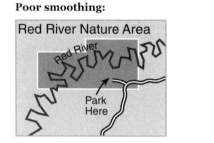

**Good smoothing:**

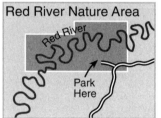

**Smoothing** is intuitive: features which are *naturally* smooth and sinuous (often natural features) are more heavily smoothed. Features which are *not* smooth (usually human features) are lightly smoothed or not smoothed at all. Computers smooth features by automatically rounding angles which exceed a set limit.

**Questions to ask when smoothing:**

- ✓ how much can you smooth a feature without losing its character?
- ✓ does smoothing a feature make it more difficult to recognize?
- ✓ does smoothing a feature make it easier to recognize?
- ✓ does the smoothing of a feature make the map less cluttered and busy-looking?

# displacement

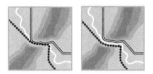

*Displacement moves features which interfere with each other apart. Moving features away from their actual location makes the features easier to differentiate and understand.*

**Map Intent:** For a state road map, where it is important for map readers to distinguish the relative location of roads, railroads, and rivers, displace the railroad and road so the relation between these features in the gap in the ridge is clear.

**Poor displacement:**

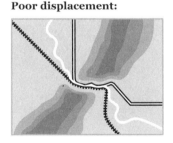

**Good displacement:**

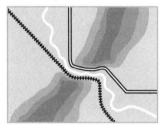

**Displacement** of map features sacrifices location accuracy for visual clarity. Displacement of point and line features is common on maps where such features are crowded together in certain areas.

**Questions to ask which guide displacement:**

     ✓ are important map features interfering with each other?
     ✓ will the slight movement of a map feature make it and its neighboring features easier to distinguish?
     ✓ will the slight movement of a map feature lead to confusion because the feature has been moved?
     ✓ does the displacement of map features help to make the map look less cluttered?

# *enhancement*

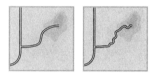

*Enhancement of map features occurs when the map maker knows enough about the feature being mapped to add details which aid in understanding the feature. A few bumps and angles, for example, may be added to a road the map maker knows is winding and angular, so the map reader understands the actual character of the road.*

**Map Intent:** It is important for a locator map that may be used by people unfamiliar with an area to communicate the character of key features. Available data of the road to Overlook Point do not convey that it is a curvy road. Enhance the road to communicate to the user that it is not a straight drive up the hill.

**Poor enhancement:**

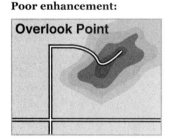

**Good enhancement:**

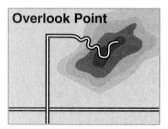

**Enhancement** *adds* detail, as opposed to *removing* detail, which most generalization procedures do. Enhancement must be used with care. The enhancement should not be deceptive, but instead should help the map reader understand important features on the map.

## Questions to ask which guide enhancement:

- ✓ do you know enough about a feature to enhance it?
- ✓ will enhancement help the map reader to better understand the feature and the map?
- ✓ does enhancing a feature make it easier to recognize?
- ✓ could enhancing a feature possibly lead to misunderstanding by the map reader?

# dimension change

*A dimension change in a feature is often necessary when changing scale and useful for removing unnecessary detail from a map. A city changes from an area to a point, a river from an area to a line. Conversely, a group of points may be transformed into an area, or a group of small areas into a larger area.*

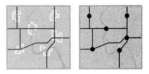

**Map Intent:** For a map of dolomite outcrops for a meeting about environmental radon, it probably is not necessary to show towns and rivers as areas. Dimension changes (areas to points and lines) makes it easier to focus on the dolomite areas.

**Poor dimensions:**

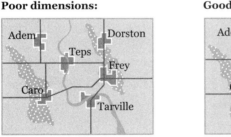

**Good dimensions:**

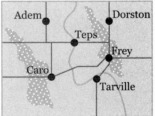

While changes in scale are the most common reason to **change dimensions,** map makers can also change dimensions of features to remove clutter and detail from a map to support its point. Computer mapping software can change feature dimensions based on the scale of the map.

**Questions to ask which guide dimension change:**

✓ would changing dimensions of a feature remove unnecessary detail?
✓ does changing the dimensions of a feature in any way affect how it is understood by the map reader?
✓ does changing the dimensions of map features help to make the map look less cluttered?

169

# Data Classification

One way maps make their point is by grouping data. The grouping, or classification, of data produces patterns. Different data classifications produce different patterns.

## qualitative classification

Beagle Great Dane Terrier

Dingo

Mutt

Terrier

Bulldog

Beagle

Chow Wolf

Hyena Chow

Beagle Retriever

Wolf

→ Wild Dogs

Feral Dogs

Domesticated Dogs

→ Canines

## quantitative classification

| Urban Population unclassified | Urban Population 5 numeric classes | | Urban Population 3 ordinal classes | |
|---|---|---|---|---|
| 23,556 | 23,556 | 18,000 - 24,000 | 18,000 - 24,000 | |
| 21,142 | 21,142 | | | |
| 18,233 | 18,233 | | | high |
| 14,199 | 14,199 | 13,000 - 15,000 | 13,000 - 15,000 | |
| 13,877 | 13,877 | | | |
| 9,676 | 9,676 | | | |
| 9,434 | 9,434 | 8,000 - 10,000 | 8,000 - 10,000 | medium |
| 8,973 | 8,973 | | | |
| 5,122 | 5,122 | | | |
| 4,889 | 4,889 | 3,000 - 6,000 | 3,000 - 6,000 | |
| 3,322 | 3,322 | | | low |
| 1,022 | 1,022 | | | |
| 988 | 988 | 600 - 2,000 | 600 - 2,000 | |
| 756 | 756 | | | |

170

# qualitative classification

**Six qualitative classes:**

**Two qualitative classes:**

# quantitative classification

**Unclassified:**

**Four classes:**

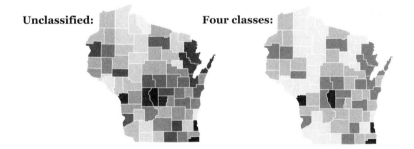

# Data Classification: Qualitative Data

Data classification is shaped by your goals for your map. In general, features in the *same* class should be more *similar* than dissimilar; features in *different* classes should be more *dissimilar* than similar. Use color *hue* and *shape* and *texture* to symbolize different classes of *qualitative* data.

## *qualitative point data*

Classification reveals patterns that are difficult to see in unclassified data. Students poll community members about social issues to learn about community politics. The bottom left classification is not very revealing. The bottom right classificiation reveals more about the political landscape. Include the unclassified data so map viewers can decide if your classification is justified.

**Unclassified:**

**Poor classification:**

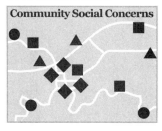

● **Family Values**

■ **Religious Values**

▲ **Legal Values**

◆ **Social Welfare Issues**

**Good classification:**

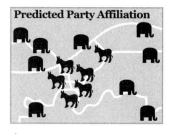

 **Democrat**

**Republican**

# *qualitative line data*

*Roads are often classified in terms of who builds and maintains them (federal, state, local). However, this classification is not the best if your map is for tourists. Your goal for the map (tourism) should shape classification (tourism-based classes of roads).*

**Unclassified:**

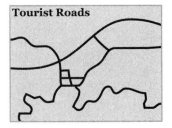

**Poor classification:**

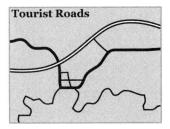

Federal Highway
State Highway
Local Road

**Good Classification:**

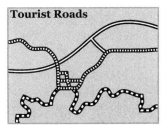

Fast Travel
Historical Sites
Natural Beauty

# qualitative area data

*Criteria external to data can serve as the basis of classification. When classifying soils, geology, or land-use data, for example, follow professional or academic classification standards. If standards vary, note which standards you chose on the map. A map created for a city planning department should use the department's land-use classification. While the map on the left may be OK if you are looking for an apartment or house, the map on the right follows land-use classification conventions.*

**Poor classification:**

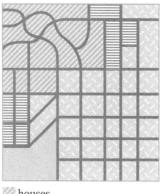

▨ houses
≣ apartments
▨ companies

**Good classification:**

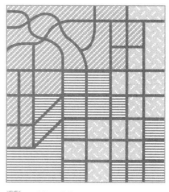

▨ residential
≣ industrial
▨ commercial

*Be aware that most qualitative data on maps has been classified. Attempt to determine the criteria upon which the classification decisions were made. Be wary of maps with no logical criteria. Conversely, a map may classify data based upon criteria suitable for one purpose but not necessarily for others.*

**USGS 1:100,000 Topographic Map**

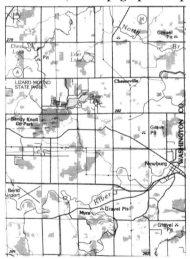

*U.S. Geologic Survey (USGS) topographic maps classify vegetation into two categories: vegetation (grey areas on this map) or no vegetation. A user of this map may believe the classification to be based on* **ecological criteria.**

*However, vegetative areas are classified based on* **military criteria:** *vegetative areas are areas with tree cover at least 6 feet tall which can hide military troops. It is a classification for army guys, not ecologists!*

175

# Data Classification: Quantitative Data

Quantitative data classification requires decisions about the number of classes and class boundaries. Quantitative data classification is shaped by external criteria or by the characteristics of the data themselves. Use color *value* and *size* to symbolize different classes of *quantitative* data.

## *quantitative point data*

Data classification generates spatial patterns in the data (which may not be evident in the unclassified data). For a map created for a community meeting, the classification on the right (bottom) is better, as it shows what is most important: whether the well water is safe for humans or not.

**Unclassified:**

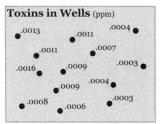

**Toxins in Wells** (ppm)

.0013  .0011  .0004
.0011  .0007
.0016  .0009  .0003
.0009  .0004
.0008  .0006  .0003

**Poor classification:**

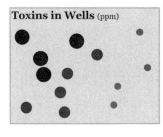

**Toxins in Wells** (ppm)

⬤ over .0010

🔴 .0006 to .0009

• under .0005

**Good classification:**

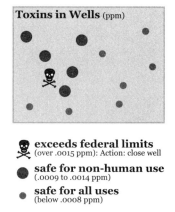

**Toxins in Wells** (ppm)

☠ **exceeds federal limits**
(over .0015 ppm): Action: close well

⬤ **safe for non-human use**
(.0009 to .0014 ppm)

• **safe for all uses**
(below .0008 ppm)

# *quantitative line data*

A map intended to help guide the restructuring
of police patrol routes should classify data,
in this case average vehicle speeds, in
categories appropriate to the task: increase,
maintain, or decrease patrols (suggesting a
three-class map).

**Unclassified:**

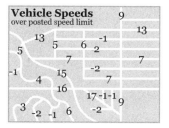

**Poor classification:**

**Amount over posted speed**

**over 15 mph**
Action: increase patrols

**11 to 14 mph**
Action: increase patrols

**5 to 10 mph**
Action: maintain patrols

**0 to 4 mph**
Action: maintain patrols

**below 0 mph**
Action: decrease patrols

**Good Classification:**

**Amount over posted speed**

**over 10 mph**
Action: increase patrols

**0 to 9 mph**
Action: maintain patrols

**under posted speed**
Action: decrease patrols

# quantitative area data

*You must decide the **number of classes.***
*Fewer classes often result in distinct patterns;*
*more classes often result in complex patterns.*
*Which option is best depends on why you are*
*making the map.  This map shows the density*
*of **mobile homes** (**dark** = higher density).*

**1 class:**

**2 class:**

**3 class**

**4 class:**

**6 class:**

**8 class:**

**12 class:**          **Unclassified:**

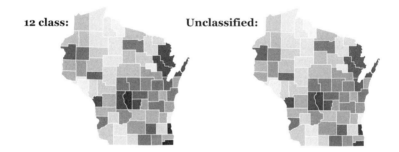

When choosing the number of classes consider that:

- ✓ A 2 class map can be suitable for binary (yes|no) data or data with negative and positive values.

- ✓ 4 to 8 classes usually ensures that map readers can see distinct patterns and match a particular shading on the map to the legend. A good default.

- ✓ More than 8 classes will produce more complex patterns, but map readers may not be able to match a particular shading on the map to the legend.

- ✓ Unclassified data (each area has a unique shading corresponding to its unique value) produce the most complex and ungeneralized patterns.

- ✓ Classify your data using different numbers of classes, and look at how patterns in your data vary. *Think about your data and your goals for your map, and make an intelligent decision.*

# quantitative area data

*In addition to determining the number of classes, you must decide where to place **boundaries between the classes.** Classification schemes set these boundaries. This map shows the density of **mobile homes** (**dark** = higher density) unclassified and as a 5-class map using four different classification schemes.*

**Unclassified:**

**Equal Interval:**

**Quantile:**

**Natural Breaks:**

**Unique:**

# thinking drives classification

*Poverty is a contentious issue. Debates rage over defining poverty, why it exists, and how to address it. The U.S. Census Bureau provides official data on poverty in the U.S., and different classifications of Census 2000 poverty data follow.*

*It is easy to get the percent of people in each county in the U.S. who live in a state of official poverty. But choosing how to map the data is not as easy. Common (and equally valid) data classification schemes – methods for placing boundaries between the classes on a map – are easy to generate but difficult to choose from. Understanding the benefits and problems with each classification scheme is vital, as is clarifying why you are making the map. Together, these guide the thinking behind choosing the most appropriate classification scheme for your data.*

## graphing data

*Selecting a classification scheme without examining your data as a graph is a bad idea. As examples in this section reveal, classification schemes can mask important characteristics of your data and perhaps undermine the goal of your map.*
*A simple histogram can be constructed from your data: the x-axis is your data variable (from low to high) and the y-axis the number of occurrences of each value:*

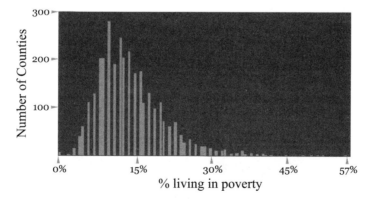

*The 2000 U.S. poverty data have a cluster of counties near the lower to mid-end of the graph, with a smaller number of counties skewed out to 57%. You can easily note where a classification scheme places class boundaries, which values are grouped together, and which values are in different groups. If a particular classification scheme seems to violate the basic classification rule (features in the same class should be more similar than dissimilar; features in different classes should be more dissimilar than similar), then consider a different classification scheme. Consider placing the graph on your final map, so map users can see how the data are classified.*

181

# unclassified scheme

To create an **unclassified scheme** assign a unique visual shade to every unique data value. In essence, **each unique data value is in its own class.** Unclassified schemes produce complex, highly variable, and subtle patterns by minimizing the amount of data generalization and simplification.

This map, due in large part to the concentration of counties near the lower end of the range of values, suggests that poverty is not a significant issue in most places, that the number of people living in poverty is somewhat similar across the U.S., and that there are few places with very high poverty.

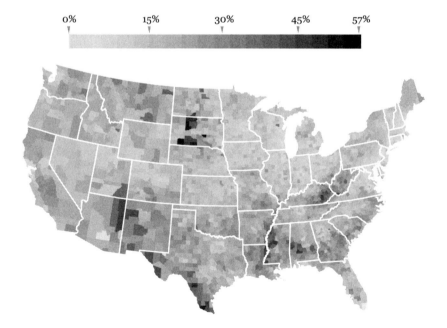

# quantile scheme

**Quantile schemes place the same number of data values in each class.** *Quantile schemes are attractive in that they always produce distinct map patterns: a quantile classification will* **never** *have empty classes, or classes with only a few or too many values. Quantile schemes look great. The problem with quantile schemes is that they often place similar values in different classes or very different values in the same class.*

*The map suggests that poverty is a significant issue in many counties, and the numerous counties in the top, darkest classes impart a rather ominous view of poverty in the United States.*

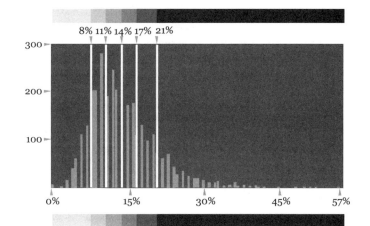

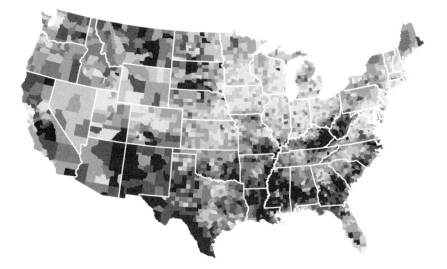

# equal-interval scheme

**Equal-interval schemes place boundaries between classes at regular (equal) intervals.** *Equal-interval schemes are easily interpreted by map readers, and are particularly useful for comparing a series of maps (which necessitates a common classification scheme). Equal-interval schemes do not account for data distribution, and may result in most data values falling into one or two classes, or classes with no values.*

*The map suggests that poverty is not an issue in most places, as there are few counties in the highest three classes.*

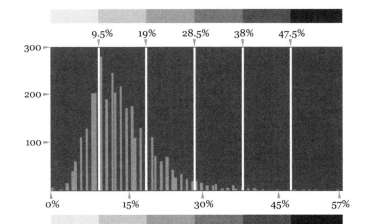

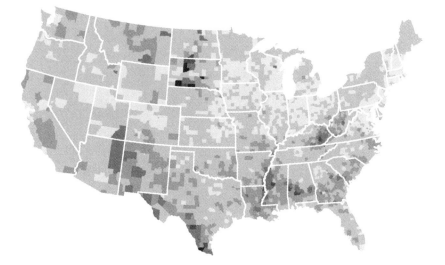

# natural-breaks scheme

**Natural-breaks schemes minimize differences between values within classes and maximize differences between values in different classes.** *Class boundaries can be determined subjectively, by graphing data and looking for "natural breaks" in the data distribution. Natural breaks are determined by algorithms which seek statistically significant groupings in a set of data. Natural-breaks schemes can serve as a default classification scheme, as they take careful account of the characteristics of the data distribution.*

*The map makes poverty seem more significant than the equal-interval map does, but is not quite as ominous as the quantile scheme.*

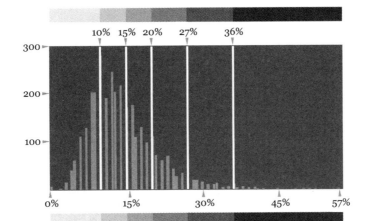

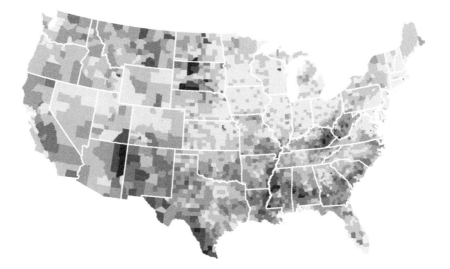

# unique scheme

**Class boundaries sometimes need to be set in accordance with external criteria:** *immigrant population by state is classified in terms of federal requirements for applying for immigrant-related grants, for example. Unique schemes require an understanding of the broader context of the data.*

*The map below is made for a study of counties with very high poverty. The researchers are not interested in any county with less than 25% of the population in poverty. The rest of the data are divided into groups which will aid in the analysis of high-poverty counties. While excellent for the study, this map and classification is not good in general, as it suggests poverty is isolated and rare.*

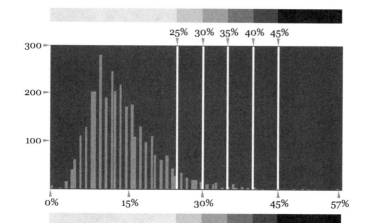

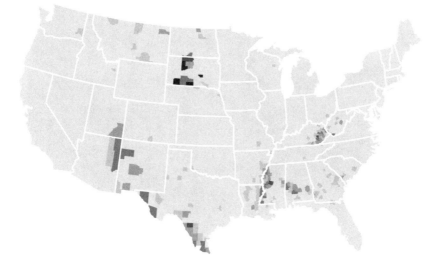

**Think about what you are seeing:** *Looking at a number of schemes brings to the foreground geographic facts, both the facts that are variously emphasized and those that are preserved through every variation.*

*The unique scheme displays counties with the highest rates of poverty. While this can be seen in every scheme, the unique scheme isolates, and so draws attention to, regions of the country where poverty has been a long-standing reality: the coal-mining heart of Appalachia; the old "Cotton Belt," specifically those counties with the highest proportion of slaves at the time the Civil War began; the counties in the Rio Grande Valley, where the population is overwhelmingly Hispanic; and the reservations of the Navajo, Lakota, Sioux, and other Native Americans. The unique scheme picked out the regions of historically outstanding social injustice.*

*Unless you looked at a lot of maps, you might not have identified these regions as anything other than those with high levels of poverty. It takes many different kinds of maps to begin to make sense of the world.*

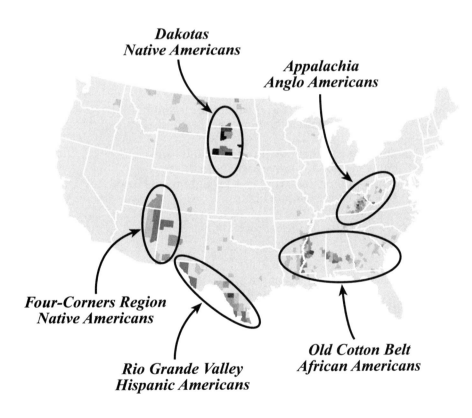

**Dakotas**
**Native Americans**

**Appalachia**
**Anglo Americans**

**Four-Corners Region**
**Native Americans**

**Rio Grande Valley**
**Hispanic Americans**

**Old Cotton Belt**
**African Americans**

"That's a fine map," said Big Tiger. "It's useful to be able to look up beforehand the places we reach later."

"Are there really bandits about here?" asked Christian.

"Perhaps it's written on the map," Big Tiger ventured. "Look and see."

Fritz Muhlenweg. *Big Tiger and Christian.* (1952)


He had brought a large map representing the sea
      without the least vestige of land;
and the crew were much pleased when they found it to be
      a map they could all understand.

Other maps are such shapes with their islands and capes,
      but we've got our brave captain to thank
(so the crew would protest) that he's brought us the best,
      a perfect and absolute blank.

Lewis Carroll. *The Hunting of the Snark.* (1876)


Airey's Railway Map is almost unique in its way, devoting itself to its subject with a singleness of purpose which is really almost sublime, and absolutely ignoring all such minor features of the country it portrays as hills, roads, streets, churches, public buildings, and so forth. It is rather startling at first to find the Metropolitan Railway pursuing its course through a country as absolutely devoid of feature as was the "Great Sahara" in the good old African maps...

Charles Dickens. *London Guide.* (1879)


188

# more information...

Mark Monmonier includes a succinct discussion of map generalization in *How to Lie with Maps*, 2nd ed. (University of Chicago Press, 1996). An excellent source for data classification is Terry Slocum et al., *Thematic Cartography and Visualization* (Prentice Hall, 2003). The section "Visual and Statistical Thinking: Displays of Evidence for Making Decisions" from Edward Tufte's *Visual Explanations* (Graphics Press, 1997) dramatizes the power of thinking visually about statistics.

Two excellent guides to graphs are William Cleveland's *The Elements of Graphing Data* (Wadsworth, 1985) and Stephen Kosslyn's *Elements of Graph Design* (Freeman, 1994).

In *Information Anxiety* (Bantam, 1989) Richard Wurman illustrates classification graphically in three pages about the 130 breeds of dogs recognized by the American Kennel Club: well worth taking a look at.

A curious tale about what happens when you *don't* generalize is Jorge Luis Borges' short story "Funes the Memorius" (in *Labyrinths: Selected Stories and Other Writings,* New Directions, 1964).

**Sources:** Bill Bunge's "Continents and Islands of Mankind" is re-created from his *Field Notes: Discussion Paper No. 2* (self published, no date). Much of this chapter is based on Dent (1998) and Slocum et al. (2003).

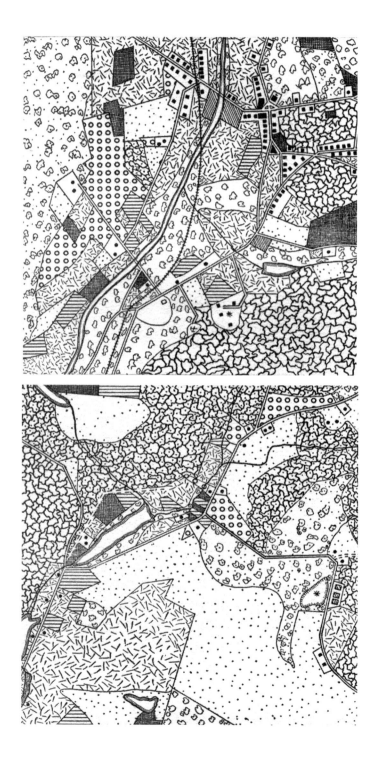

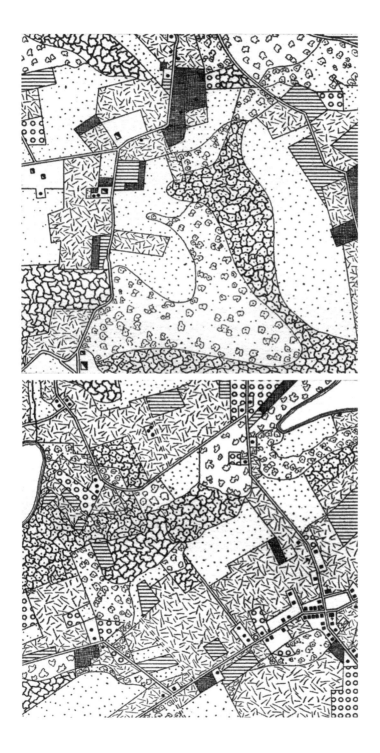

Small Muslim
Shrine or Tomb

Gantry
Crane

Cave

Patrol
Route

Manholes along
Sewer line

Demolished
Building

Oil
Seepage

Pigeon
Roost

Forest Ranger's
House

Train Roundhouse
and Track

Thorny
Scrub

Stone
Mosque

Dry
Dock

Shipwreck

Fenced
Fuel Tanks

Fishing
Stakes

Fir Tree
Landmark

Semaphore
Signal

Essential
Oil Plants

Common
Grave

Tank
Trap

Whirlpool

Corduroy
Road

Trigonometrical
Point on
Burial Mound

Fog
Station

Gas Drip
Manhole

Cairn

Ruins

Stone or Brick
Wall

Bunker

Log
Chute

Moss and
Grass

Railroad
Ferry

Birch
Grove

Pine or Cedar
Landmark

Casuarina

Winter
Road

Overland Oil
Pipe & Station

Marabout

Cattle
Run

Wooden
Windmill

Post Relay
Center

Forest Service
Station

Quarry

Prospect
Well

Radio
Station

Kilometer
Post

Narrow Gauge
Railroad

Tent

Windfall

Oil
Derrick

Rail or
Wattle Fence

Trench,
Communication Trench,
& Dugout

Flour
Mill

Pit

Stone
Church

Rail
Lubricators

Driftwood
Accumulation

Larch Tree
Landmark

Takyrs
(salt clay flats)

Polar
Station

Slipway

Volcanic
Manifestation

Forest
Boundary

Hedge

Search-
light

Fireproof
Building

Cotton

Explosives
Area

Vegetable
Storage

Dismantled
Railroad

Artificial
Embankment

Saw
Mill

Burial
Mound

Cemetery
with Trees

Nine

Fire Hydrants
along water line

Statue or
Monument

Emergency
Landing Field

Fenced
Apiary

# How do you make data into visual marks?

Quarantine

Metal
Fence

Power
Station

Cattle Burying
Ground

Patrol
Look-out

Leased Land
Boundary

Cave

Small Muslim Shrine or Tomb

Patrol Route

Gantry Crane

Manholes along Sewer line

Oil Seepage

Pigeon Roost

Demolished Building

Train Roundhouse and Track

Thorny Scrub

Stone Mosque

Forest Ranger's House

Dry Dock

Fishing Stakes

Fenced Fuel Tanks

Shipwreck

Fir Tree Landmark

Semaphore Signal

Common Grave

Essential Oil Plants

Whirlpool

Tank Trap

Gas Drip Manhole

Cairn

Corduroy Road

Trigonometrical Point on Burial Mound

Fog Station

Bunker

Log Chute

Stone or Brick Wall

Ruins

Pine or Cedar Landmark

Birch Grove

Moss and Grass

Railroad Ferry

Casuarina

Winter Road

Marabout

Cattle Run

Overland Oil Pipe & Station

Wooden Windmill

Post Relay Center

Quarry

Prospect Well

Radio Station

Forest Service Station

Kilometer Post

Narrow Gauge Railroad

Tent

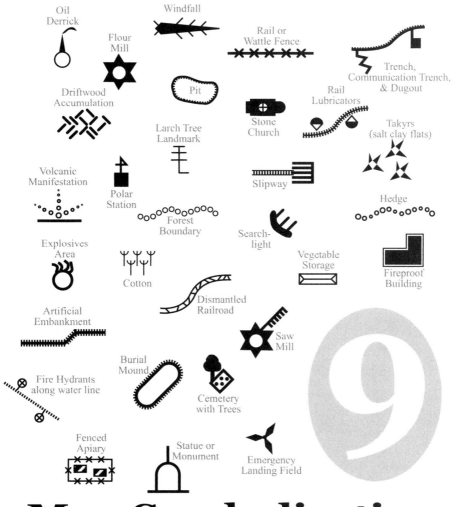

Oil Derrick

Flour Mill

Windfall

Rail or Wattle Fence

Trench, Communication Trench, & Dugout

Driftwood Accumulation

Pit

Rail Lubricators

Stone Church

Takyrs (salt clay flats)

Larch Tree Landmark

Volcanic Manifestation

Polar Station

Slipway

Hedge

Explosives Area

Forest Boundary

Search-light

Vegetable Storage

Fireproof Building

Cotton

Dismantled Railroad

Artificial Embankment

Saw Mill

Fire Hydrants along water line

Burial Mound

Cemetery with Trees

Fenced Apiary

Statue or Monument

Emergency Landing Field

# Map Symbolization

***Map symbols allow us to put just about anything on a map.***
These symbols are from many old sources, including 1950s Soviet
topographic maps, the U.S. Coast and Geodetic Survey, the Federal
Board of Surveys and Maps, U.S. Forest Service maps, Cleveland Regional
Underground Survey maps, The American Railway Engineering
Association, and the U.S. War Department.

Quarantine

Metal Fence

Power Station

Cattle Burying Ground

Patrol Look-out

Leased Land Boundary

# Map Symbolization

Everything on a map is a symbol. A map symbol is a visual mark systematically linked to the data and concepts shown on a map. Consciously and critically choose symbols that help your map do what you want it to do.

## *symbol by relationship*

Some map symbols *intuitively* suggest general kinds of data.

A map showing the population of different cities uses circle sizes from small to large: sizes vary in *amount,* as do the data.

A map showing restaurants, antique stores, and museums in a town uses different shapes: shapes vary in *kind,* as do the data.

## *symbol by resemblance*

Other map symbols *look* like particular data or concepts.

A map showing the location of airports uses an airplane symbol for airports: the symbol *looks* like an airplane, and is associated with an airport.

Maps in a war atlas use red symbols to show the location of battles: the symbol *looks* like an explosion, and is associated with a battle. People often associate red with danger or conflict.

## *symbol by convention*

Some symbols "make sense" even though they may not entirely make sense.

The U.S. Geological Survey uses the Christian cross to symbolize all places of worship - church, mosque, synagogue. We know what they mean, though it's not very politically correct.

A map showing the earth's oceans uses blue for water. But is water out in the real world blue? Not usually. Blue on a map, however, suggests water - it's a *convention.* If you depart from conventions (color your oceans their actual color) you could confuse your map's readers.

Basic **map symbolization** issues include:

**1** **Different kinds of data:** data at points, along lines, and in areas; differences in kind and number; individual or aggregate data.

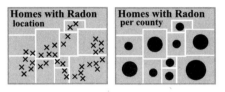

**2** **The visual variables:** choosing map symbols.

The visual variable **size** readily implies...

...variation in **quantity**.

The visual variable **shape** readily implies...

...variation in **quality**.

**3** **Symbolizing aggregate data:** guidelines for choosing among choropleth, graduated symbol, dot, and surface maps.

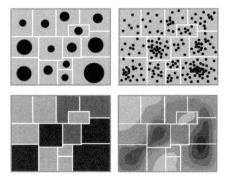

# 1 Different kinds of data

Mappable data vary immensely. To assist in symbolizing your data, it is helpful to ask three questions: Are your data at **points,** along **lines,** or in **areas**? Are your data **qualitative** (vary in kind) or **quantitative** (vary in number)? Do you have **individual** data values, or are they **aggregated** (grouped)?

## *points, lines, or areas?*

The first step in choosing appropriate symbols for your map is to consider the dimensions of the data you are mapping. Most mappable data are at **points** (zero dimensions), **lines** (one dimension), or in **areas** (two dimensions).

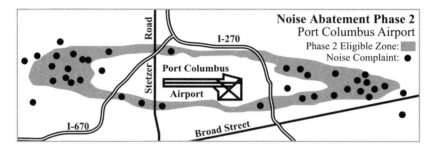

Data at **points** include

- ✓ location of an airport noise complaint.
- ✓ location of badger burrows in a park.
- ✓ location of car accidents in a county.

Data along **lines** include

- ✓ roads and airport runways.
- ✓ rivers.
- ✓ route of bicycle race.

Data in **areas** include

- ✓ airport noise abatement zone.
- ✓ size and shape of a park.
- ✓ soil types.

## *things are complicated!*

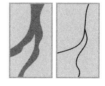

The dimensionality of the same data may **vary** due to scale or your choices as a map maker: a river, for example, can be shown as an *area* or a *line*.

# qualitative or quantitative?

Consider next whether your data vary in either **quality** (differences in kind) or **quantity** (differences in amount). Some data are not easily qualitative or quantitative.

## Qualitative Data

- ✓ house location.
- ✓ border or boundary.
- ✓ land vs. water.
- ✓ religious denominations.
- ✓ animal species.
- ✓ census racial types.
- ✓ house styles.
- ✓ cheese species.
- ✓ plant types.
- ✓ musical tastes.
- ✓ sexual orientation.
- ✓ sports team affiliations.
- ✓ political affiliation.

## Quantitative Data

*Ordinal data:* distinctions of order but with no measurable difference between the ordered data.

- ✓ high, medium, low risk zones.
- ✓ acceptable & unacceptable noise levels.
- ✓ restaurant ratings (1, 2, 3 stars).

*Interval data:* distinctions of order with measurable differences between the ordered data, but no absolute zero.

- ✓ temperature Fahrenheit or Centigrade.

*Ratio data:* distinctions of order with measurable differences between the ordered data and an absolute zero.

- ✓ population in countries.
- ✓ murder rate per country.
- ✓ temperature Kelvin.

# things are complicated!

The *shape* of Wisconsin seems purely *qualitative.* But shape has quantitative aspects: the number of sides, ratios of angles, and size compared to other states.

A map showing the *number* of farms in each Wisconsin county seems purely *quantitative* (more farms = larger circle), but the circles themselves are qualitative, as is the arrangement, suggesting the shape of the state.

# *individual or aggregate?*

When Abraham Verghese made maps of his AIDS patients in the 1980s, he had data about where they currently lived, and where they had lived in the preceding decade.

His data consisted of infected individuals (facts) at individual addresses (locations). These are **individual data.**

This map shows where patients of Dr. Verghese lived prior to moving home.

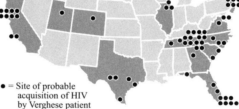

● = Site of probable acquisition of HIV by Verghese patient

Dr. Verghese made maps to help him understand why he had so many AIDS patients in his rural practice. Other researchers were making maps of AIDS in the 1980s using aggregate data. To produce such a map one has to sum all the individual cases in a given areal unit (grouped, or aggregated, facts), often in counties (locations). These are **aggregate data.** Such a map is useful to understand the prevalence of AIDS and its spread.

Darker = more on this map of total AIDS cases in 1986. This map reveals similar patterns to Dr. Verghese's map.

# **2** The visual variables

Particular visual variables intuitively suggest important characteristics of your data.

✓ If your data are **qualitative,** choose a visual variable which suggests **qualitative** differences (shape, color hue).

✓ If your data are **quantitative**, choose a visual variable which suggests **quantitative** differences (size, color value).

✓ Some visual variables can be manipulated to suggest either qualitative or quantitative differences (texture).

| | *Points* | *Lines* | *Areas* | *Best to show* |
|---|---|---|---|---|
| *Shape* | | *possible, but too weird to show* | *cartogram* | *qualitative differences* |
| *Size* | | | *cartogram* | *quantitative differences* |
| *Color Hue* | | | | *qualitative differences* |
| *Color Value* | | | | *quantitative differences* |
| *Color Intensity* | | | | *qualitative differences* |
| *Texture* | | | | *qualitative & quantitative differences* |

Map symbolization is a complex process. The visual variables can guide basic map symbolization decisions, but they cannot solve all map symbolization issues.

A review of the basic visual variables with examples follows...

**Visual variables described and illustrated**

Understanding the nature of visual variables will
help you select map symbols which clearly make
the point you want your map to make.

## shape

*Map symbols with different shapes intuitively
imply differences in quality. A square is not
more or less than a circle, but is different in
kind. Map symbol shapes can be pictorial or
abstract.*

### Poor use of shape:

**Active Hate Groups, 2003**

★ 34 - 54
♦ 22 - 33
● 11 - 21
■ 4 - 10
▲ 0 - 3

▲ Alaska
▲ Hawaii

**Shape is a poor choice for
showing quantitative data**

✓ using shape makes it difficult
to see the patterns on the
map, as the symbols do not
suggest the order (low to
high) in the data.

### Good use of shape:

**Shape is a good choice for
showing qualitative data**

✓ different shapes suggest the
qualitatively different groups.

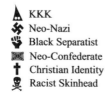 KKK
Neo-Nazi
Black Separatist
Neo-Confederate
† Christian Identity
Racist Skinhead

**Dominant Hate Group, 2003**

Alaska
Hawaii

# size

*Map symbols with different sizes intuitively imply differences in quantity. A larger square implies greater quantity than a smaller square.*

## Good use of size:

### Active Hate Groups, 2003

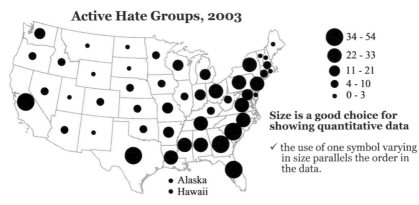

● 34 - 54
● 22 - 33
● 11 - 21
● 4 - 10
• 0 - 3

• Alaska
• Hawaii

**Size is a good choice for showing quantitative data**

✓ the use of one symbol varying in size parallels the order in the data.

## Poor use of size:

**Size is a poor choice for showing qualitative data**

✓ different sizes suggest order in the data, rather than the qualitatively different groups.

• KKK
• Neo-Nazi
● Black Sepratist
● Neo-Confederate
● Christian Identity
● Racist Skinhead

### Dominant Hate Group, 2003

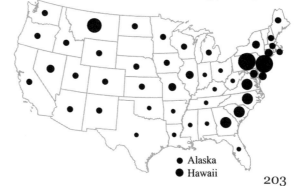

• Alaska
• Hawaii

203

# color hue

*Color hue refers to different colors such as red and green. Symbols with different hues readily imply differences in quality. Red is not more or less than green, but is different in kind.*

## Poor use of color hue:

### Iraq War Casualties
#### As Percent of State Population: Aug 1, 2004

- 0.0000060% - 0.0000150%
- 0.0000042% - 0.0000059%
- 0.0000031% - 0.0000041%
- 0.0000021% - 0.0000030%
- 0.0000008% - 0.0000020%

### Hue is a poor choice for showing quantitative data

✓ using hue makes it difficult to see the patterns on the map, as the colors do not suggest the order (low to high) in the data.

## Good use of color hue:

### Color hue is a good choice for showing qualitative data

✓ qualitatively different hues parallel the qualitatively different data.

- Bush Win
- Kerry Win

### Presidential Election, 2004

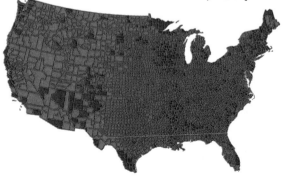

# color value

*Color value refers to different shades of one hue, such as dark and light red. Map symbols with different values readily imply differences in quantity. Dark red is more than light red.*

## Good use of color value:

### Iraq War Casualties
#### As Percent of State Population: Aug 1, 2004

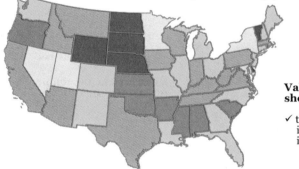

| | |
|---|---|
| ■ | 0.0000060% - 0.0000150% |
| ■ | 0.0000042% - 0.0000059% |
| ■ | 0.0000031% - 0.0000041% |
| ■ | 0.0000021% - 0.0000030% |
| ■ | 0.0000008% - 0.0000020% |

**Value is a good choice for showing quantitative data**

✓ the use of one hue varying in value parallels the order in the data.

## Poor use of color value:

### Presidential Election, 2004

**Color value is a poor choice for showing qualitative data**

✓ values suggest an ordered difference, which is not appropriate for these data.

▨ Bush Win
▨ Kerry Win

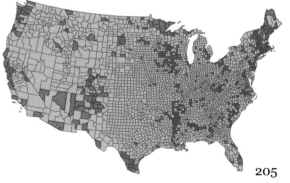

# color intensity

*Color intensity (or saturation) is a subtle visual variable that is best used to show subtle data variations, such as binary (yes / no) data that are difficult to categorize as qualitative or quantitative.*

## Poor use of color intensity:

### Hate Crimes, 2002
#### As Percent of State Population

| | |
|---|---|
| ■ | 0.000050% - 0.000069% |
| ■ | 0.000035% - 0.000049% |
| ■ | 0.000021% - 0.000034% |
| ■ | 0.000009% - 0.000020% |
| ■ | 0.000000% - 0.000008% |

**Intensity is a poor choice for showing quantitative data**

✓ Intensity may suggest order, but due to the lack of variation in value the sense of order is weak.

## Good use of color intensity:

### African American Presence, 2000

**Intensity is a good choice for showing yes/no data**

✓ intensity, like binary yes or no data, is neither qualitative nor quantitative.

| | |
|---|---|
| ■ | No African Americans |
| ■ | One or More African Americans |

# *texture*

Texture (or pattern) variations can imply both qualitative (brick vs. cloth) and quantitative (coarse vs. fine) differences. Select textures so that they suggest the qualitative or quantitative character of your data. Texture is useful for showing qualitative differences between areas on black-and-white maps.

## Poor use of texture:

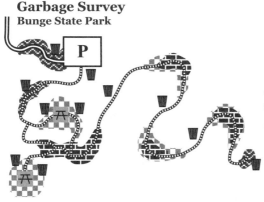

### Garbage Survey
**Bunge State Park**

▨ Cigarette Butts
▧ Paper Debris
▦ Glass & Cans

**Textures can be visually noisy and imply ordered differences**

✓ Be careful with textures that look like something: glass and cans are shown with a brick pattern, which does not make sense. The patterns here are noisy and interfere with the trail pattern.

## Good use of texture:

### Garbage Survey
**Bunge State Park**

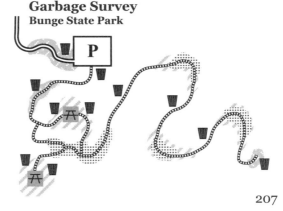

**Texture can be good for showing qualitative data**

✓ Select textures that are not visually noisy and that suggest the qualitative differences in the data.

▨ Cigarette Butts
▧ Paper Debris
▦ Glass & Cans

207

# bivariate

*Two or more visual variables can be used to symbolize the two or more data variables. Bivariate symbols allow you to add more information to your map.*

# Hate Groups and Hate Crimes

○ Alaska
○ Hawaii

**Total Active Hate Groups, 2003**

- ● 34 - 54
- ● 22 - 33
- ● 11 - 21
- ● 4 - 10
- • 0 - 3

**Hate Crimes, 2002**
**As Percent of State Population**

- 0.000050% - 0.000069%
- 0.000035% - 0.000049%
- 0.000021% - 0.000034%
- 0.000009% - 0.000020%
- 0.000000% - 0.000008%

# Crestview Alley Debris Survey

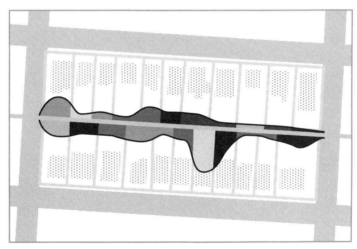

**Size and value represent two variables of data**

✓ Size (width) of line represents amount of organic debris and garbage adjacent to the alley. The further the line is away from the alley, the more debris and garbage.

✓ Value of line represents property value (dark means more).

# ③Symbolizing aggregate data

Symbolizing data that have been grouped into geographic areas (countries, states, etc.) is complicated: the same data can be mapped as a choropleth, graduated symbol, dot, or surface map. Choosing among these methods requires understanding what you are mapping and your goals for the map.

## Incidence of AIDS in Pennsylvania, 1995

One of the most common types of mappable data is associated with geographic areas: *one* data value is associated with each geographic area. In the set of maps which follow, each county in Pennsylvania has one value, representing total AIDS cases: the same data are mapped with four methods. AIDS is a contagious disease unevenly spread through geographic space. Available AIDS *data* usually consist of a single value for a county, state, or country.

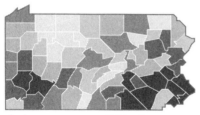

*darker means more, lighter less*

A **choropleth** map varies the shading of each area in tandem with the data value associated with it. On this map, dark is more and light is fewer AIDS cases.

This map suggests that the incidence of AIDS is uniform throughout each county, with potentially abrupt changes at county boundaries. This map is useful for an AIDS educator working at the county level, suggesting that AIDS is everywhere in the county, and everyone must take precautions, or from county representatives seeking aid for AIDS care from a legislature.

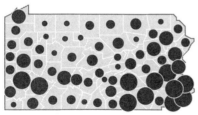

*larger circles mean more, smaller less*

A **graduated symbol** map varies the size of a symbol centered on each area in tandem with the data value associated with it. On this map, large is more and small is fewer AIDS cases.

This map suggests a single data value for each county. This map does *not* suggest a uniform distribution of AIDS cases within each county. This map is useful for the official yearly health report for the State of Pennsylvania. The map indicates that there are AIDS cases, but the individual symbols subtly suggest AIDS is contained and under control.

210

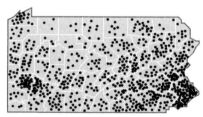

*density of dots reveals density of cases*

A **dot** map varies the number of dots in each area in tandem with the data value associated with it. On this map, one dot equals 30 AIDS cases. The location of dots *does not* show the specific location of AIDS cases; rather, the density of dots in each area represents more or fewer AIDS cases in the area.

This map suggests specific cases of AIDS, and that AIDS cases are spread throughout each county in an uneven manner. This map is useful for a epidemiologist presenting to colleagues at a professional meeting.

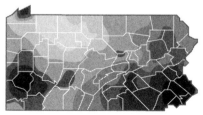

*darker means more, lighter less*

The **surface** map creates an abstract surface based on the single data value associated with each area. Imagine a pin stuck in each county, where the height of the pin varies with the number of AIDS cases. Now imagine a surface created by a sheet laid over these pins. On this map, darker (higher) areas have more and lighter (lower) areas have fewer AIDS cases.

This map suggests that AIDS is everywhere in the *state* and that AIDS is highly contagious. This map is useful for a statewide AIDS awareness campaign intended to scare teens.

211

# 3a Symbolizing aggregate data: choropleth maps

The **choropleth map** is one of the most common mapping techniques for data grouped into areas - counties, provinces, states, countries, etc. Choropleth maps vary the shading of each area along with the data value associated with the area.

## *appropriate data*

✓ *derived data (densities, rates), sometimes totals*

Mapping *total numbers* with a choropleth map is often *not* recommended, particularly when the areas on the map vary in size. A large area may have more people simply because it covers a larger area.

If you map *totals* (bold, below) and classify the data, an area with 100 people will likely be in a *different* class than the area with 500 people. The visual difference between the areas is the result of the *unequal size of the areas.*

Mapping *derived data,* like *population per square mile,* takes into account the varying size of areas on the map.

If you map *densities* (bold, below) and classify the data, both areas have 10 people per square mile and will be in the *same* class.

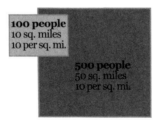

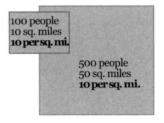

***Mapping totals with a choropleth map can be OK!***
A marketing company maps the total number of Polish-speaking U.S. citizens by county in the U.S., to assist in a plan to market Polish greeting cards. In this case, what is most important is where the most Polish-speaking folks are. But do consider a graduated symbol map for totals: it may better serve your needs.

***Your goals for your map drive map choices!*** There is no absolute "best" kind of data for a choropleth map – independent of your goals for your map. *Be aware of the problem of mapping totals with a choropleth map,* but if your goals require totals, just do it. *Then* create a graduated symbol map of the same data, and compare the maps. Please use your brain when making maps.

# *design issues* *shading*

**Poor value & legend:**

## Wisconsin Farm Density

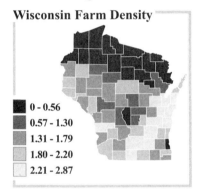

| | |
|---|---|
| ■ | 0 - 0.56 |
| ▓ | 0.57 - 1.30 |
| ▒ | 1.31 - 1.79 |
| ░ | 1.80 - 2.20 |
| □ | 2.21 - 2.87 |

✓ dark means less is not intuitive.
✓ smaller values at *top* of legend
   are not intuitive.

**Poor value & boundaries:**

## Wisconsin Farm Density

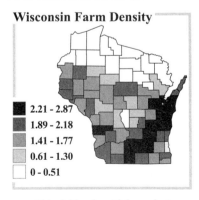

| | |
|---|---|
| ■ | 2.21 - 2.87 |
| ▓ | 1.89 - 2.18 |
| ▒ | 1.41 - 1.77 |
| ░ | 0.61 - 1.30 |
| □ | 0 - 0.51 |

✓ black blends with boundaries.
✓ white suggests no data.
✓ boundaries stand out too much.

**Good value & legend & boundaries:**

## Wisconsin Farm Density
**Farms per
sq. mile
2000**

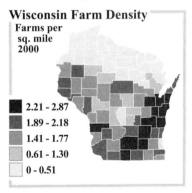

| | |
|---|---|
| ■ | 2.21 - 2.87 |
| ▓ | 1.89 - 2.18 |
| ▒ | 1.41 - 1.77 |
| ░ | 0.61 - 1.30 |
| □ | 0 - 0.51 |

✓ dark means more to most people.
✓ non-continuous legend, with larger
   values at the top and a title which
   explains the numbers in the legend.
✓ overall smoother feel without black
   and white in addition to removing
   the problems they cause.
✓ boundaries less dominant but distinct.

# Symbolizing aggregate data: graduated symbol maps

The **graduated symbol** map varies the size of a single symbol, placed within each geographic area, in tandem with the data value associated with the area.

## *appropriate data*

✓ *totals, sometimes derived data (densities, rates)*

Mapping *derived data* with a graduated symbol map is often not recommended. Graduated symbols readily imply magnitude rather than density or rates.

If you map *density* (bold, below) the symbols imply no magnitude difference in population between the two areas.

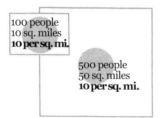

Use *totals* for graduated symbol maps. Mapping totals with graduated symbols suggests the magnitudes inherent in the data.

If you map *totals* (bold, below) the map implies a difference in total population between the two areas.

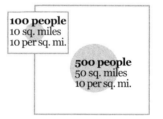

***Mapping derived data with a graduated symbol map can be OK!*** A map of global coffee consumption uses graduated coffee cups to show the percent of total global coffee consumption in each country, and relates the map to a pie chart showing the same data.

Do consider a choropleth map for this application: it may serve your needs better. But the graduated coffee cups may just be *too sweet* to pass up.

***Your goals for your map drive map choices!*** There is no absolute "best" kind of data for a graduated symbol map — independent of your goals for your map. Be aware of the problem of mapping derived data with a graduated symbol map, but if your goals require it, just do it.

# *classification*

**Classified legend:**

8001 - 10,000 persons

5001 - 8000 persons

1001 - 5000 persons

0 - 1000 persons

**Unclassified legend:**

9000 persons

6500 persons

2500 persons

500 persons

**Classified:** Use standard classification schemes: one symbol size for each class.

✓ less data detail.
✓ easier to match particular symbol to legend.
✓ easier to see distinct classes in data.

**Unclassified:** Scale each symbol to each value.  Legend includes representative symbol sizes.

✓ more data detail.
✓ harder to match particular symbol to legend.

# *design issues symbol shape*

**Squares, triangles**

✓ less compact symbol.
✓ "edgy" visual impression good for "edgy" phenomena.

**Circles**

✓ more compact symbol.
✓ smooth visual impression good for more mellow phenomena.

**Volumetric shapes**

✓ visually attractive.
✓ suggest volume phenomena.

  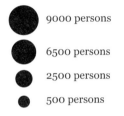

**Pictographic shapes**

✓ visually attractive.
✓ can be easy to understand.
✓ potentially too cute and thus distracting.

# *design issues* *multivariate*

**Two single-variable maps:**

**1st variable (graduated symbol):**

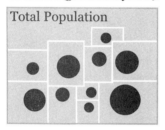

**2nd variable (choropleth):**

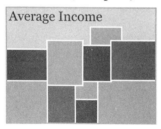

**One *bi-variate* map:**

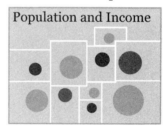

✓ more work for map user to compare two maps.
✓ harder to see correlations.
✓ less complex.
✓ uses more space.
✓ "small multiples."

✓ less work for map user to compare two maps.
✓ easier to see correlations.
✓ more complex.
✓ uses less space.

216

**Three single-variable maps:**

**1st variable (graduated symbol):**

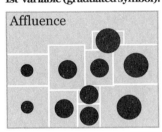
Affluence

**2nd variable (choropleth):**

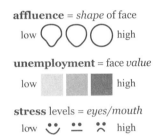
Unemployment

**3rd variable (graduated symbol):**

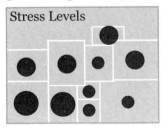
Stress Levels

**One *tri-variate* map:**

**affluence** = *shape* of face

low ◯ ◯ ◯ high

**unemployment** = face *value*

low ▢ ▢ ▢ high

**stress** levels = *eyes/mouth*

low ☺ ☷ ☹ high

**Chernoff faces (3 variables):**

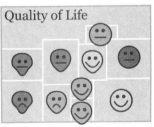
Quality of Life

✓ Chernoff faces work well for multivariate symbols, as humans are good at noting subtle differences in facial characteristics.

# Symbolizing aggregate data: cartograms

The **cartogram** is a variant of the graduated symbol map. Cartograms vary the size of *geographic areas* (rather than symbols) based on the single data value associated with the area. Cartograms, while difficult to create, are visually striking.

## appropriate data

### ✓ *totals and derived data (densities, rates)*

**Derived data:**

U.S. Suicide Rate

Cartograms of data with minimal variation from area to area can be interesting. Suicide rates (above) don't vary much around the U.S., and thus all the states are about the same size.

Cartograms are not effective when the map reader is *not familiar* with the geographic areas being varied.

**Total data:**

2004 Presidential Election: Electoral Votes

Kerry 252
Bush 286

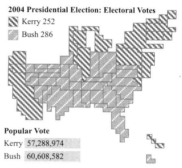

Popular Vote
Kerry 57,288,974
Bush 60,608,582

Cartograms are effective when data variation from area to area is *significant*. A cartogram of electoral votes in the U.S. removes the bias of physical geography, scaling each state to its number of electoral votes.

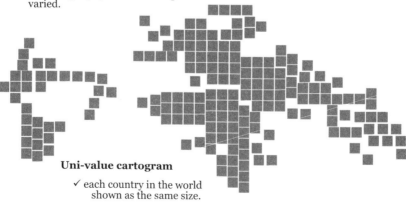

**Uni-value cartogram**

✓ each country in the world shown as the same size.

# *design issues* *cartogram form*

**Contiguous (not exploded):**

✓ adjacent areas.
✓ maintains area boundary
   relationships and adjacency.
✓ more difficult to make,
   by hand or with computers.
✓ compact.

**Non-contiguous (exploded):**

✓ non-adjacent areas.
✓ maintains area shape.
✓ easier to make (scale areas
   like graduated symbols).
✓ less compact.

# *multivariate*

**Multivariate (2 variables):**

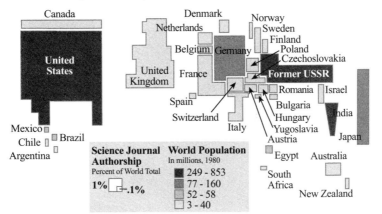

# 3d Symbolizing aggregate data: dot maps

The **dot** map varies the number of dots in each geographic area, based on the single data value associated with the area. Dots on a dot map *do not* represent the specific location of a single instance of some phenomena; rather the *density* of dots in a geographic area represents the *density* of phenomena in that area.

## *appropriate data*

### ✓ *totals, not derived data (densities, rates)*

Using a dot map for *derived data* is *not* recommended. "One dot equals 50 people per square mile" is too weird to think about.

Mapping software randomly locates dots, which is misleading. Random concentrations suggest patterns which *do not exist*. Random placement suggests concentrations of Asians in rural northern areas:

Use *totals* for dot maps. Each dot equals a number of phenomena.

Dot placement can be guided by *additional knowledge*. Mapping software may allow the use of filters to refine dot placement. Asians live mostly in urban areas in these counties. Thus *more* dots are placed near urban areas and *fewer* in rural areas:

**Poor placement:**

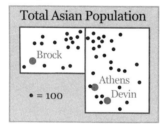

**Good placement:**

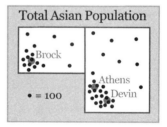

**Poor area size (random):**

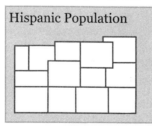

**Good area size (random)**

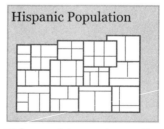

If placing dots *randomly,* the *larger* the geographic area used, the *more likely* randomly generated patterns will appear.

Make your dot map using *smaller geographic areas.* Create a dot map based on data for *townships* rather than *counties* (each county has several townships). Then remove the township boundaries.

# *classification* *dot value, size*

**Poor dot value (too dense):**

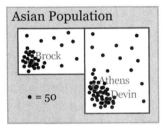

**Good dot value (change value):**

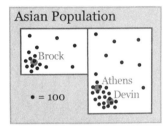

**Poor dot value (too sparse):**

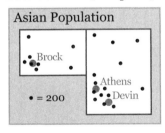

**Good dot value (change size):**

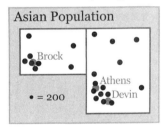

✓ adjust dot value and/or size so that dots begin to coalesce in densest areas.
✓ choose a round number for dot value (100, not 142).

# *legends*

**Fair legend:**

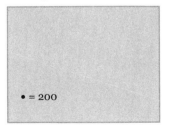

**Better legend:**

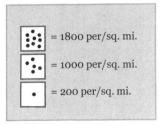

# Symbolizing aggregate data: surface maps

**Surface** maps create an abstract 3D surface (isoplethic maps) based on one data value associated with each area. Surface maps can also be created from data at points (isometric maps). Isoplethic maps, the focus of this section, suggest continuous phenomena.

## *appropriate data*

### ✓ *derived data (densities, rates)*

Using a surface (isopleth) map for *totals* is often *not* recommended, particularly when the areas on the map vary in size. A large area may have more people simply because it covers a larger area.

If you map *totals* (below) an area with 100 people will likely be at a *different* level than an area with 500 people. The visual difference between the areas is the result of the *unequal size of the areas.*

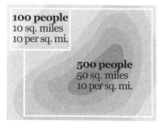

Use derived data such as *densities* for surface (isopleth) maps. Mapping people per square mile takes into account the varying size of areas on the map, allowing the map user to see real differences in the distribution of people.

If you map *densities* (below), both areas have 10 people per square mile and will be at the *same* level. The visual difference (or lack of difference) between the areas is the result of the *data*.

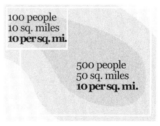

***Mapping totals with a surface map can be OK!*** A hunter wants to shoot as many coyotes as possible, due to a $20-per-head bounty. A surface map of estimated coyotes per county will guide him to where the most coyotes are. Do consider a graduated symbol map for totals: it may better serve your needs.

***Your goals for your map drive map choices!*** There is no absolute "best" kind of data for a surface map – independent of your goals for your map. *Be aware of the problem of mapping totals with a surface map,* but if your goals require totals, just do it. Then create a graduated symbol map of the same data, and compare the maps.

# *classification* *interval + scheme*

**Poor interval:**

Children per sq mi, age 0-16
U.S. Census, 2000
3 classes

The creation of surface maps from area data generates a 3D surface from a set of values. Software packages offer many different methods for interpolating this surface, and manipulation of each method's variables will lead to very different looking surfaces.

Once generated, the surface can be represented visually in different ways: filled contours (above) are common for surfaces generated using area data.

As with choropleth classification, you must choose the number of classes (here corresponding to the number of shaded levels) and the classification scheme.

Too few classes (levels), as above, will mask important variations in the abstract 3D surface.

**Good interval:**

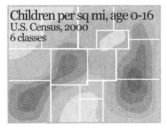

Children per sq mi, age 0-16
U.S. Census, 2000
6 classes

Increasing the number of classes (levels) from three to six reveals important variations in the abstract surface, including the location of high and low areas masked by the simpler map.

Most mapping software allows you to classify the data used for surface maps with the same methods used for choropleth maps. Many maps use an arbitrary equal-interval scheme (0 - 50, 51 - 100, etc.), and this is a good default. Custom schemes as well as natural breaks may generate more meaningful maps in some cases.

# design issues *shading*

## Fair design:

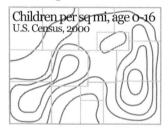

Children per sq mi, age 0-16
U.S. Census, 2000

Contour lines represent constant values slicing through the abstract 3D surface. The problem with contour lines is that they often produce a very busy map with little visual room for other data. They may also be misinterpreted as a tangible linear feature by naive map readers.

## Good design:

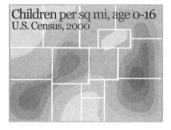

Children per sq mi, age 0-16
U.S. Census, 2000

Filled contours (above) are less busy than contour lines and allow other data to be superimposed. They also suggest a surface better than contours. A smoother surface can be approximated by more levels (classes).

# design issues *legends*

## Fair legend:

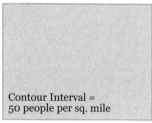

Contour Interval =
50 people per sq. mile

## Better legend:

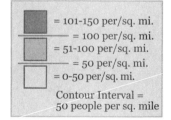

= 101-150 per/sq. mi.
= 100 per/sq. mi.
= 51-100 per/sq. mi.
= 50 per/sq. mi.
= 0-50 per/sq. mi.

Contour Interval =
50 people per sq. mile

# design issues *shading*

**Poor value:**

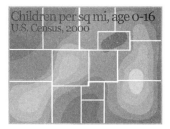

Dark tends to imply more, light less. In this case the convention is reversed: high data values are light and low values dark. This might confuse the map reader.

**Good value:**

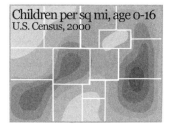

High values are dark and low values light, and the map is easier to interpret. If using color, use a range of one hue, such as light green to dark green. A spectral scheme works for temperature maps but is not a good choice for other phenomena.

# design issues *multivariate*

**One Variable:**

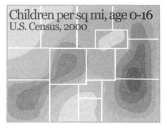

- 80 - 100 per sq. mi.
- 60 - 79 per sq. mi.
- 40 - 59 per sq. mi.
- 20 - 39 per sq. mi.
- 0 - 19 per sq. mi.

**Two Variables:**

Total Day Care Providers

50   100   10

225

"You're giving the company to a guy who thought the blue part of a map was land?"

Michael. *Arrested Development.* (TV show, 2003)

"I really can't believe it," Clevinger exclaimed to Yossarian in a voice rising and falling in protest and wonder. "It's a complete reversion to primitive superstition. They're confusing cause and effect. It makes as much sense as knocking on wood or crossing your fingers. They really believe that we wouldn't have to fly that mission tomorrow if someone would only tiptoe up to the map in the middle of the night and move the bomb line over Bologna. Can you imagine? You and I must be the only rational ones left."

In the middle of the night Yossarian knocked on wood, crossed his fingers, and tiptoed out of his tent to move the bomb line up over Bologna.

Corporal Kolodny tiptoed stealthily into Captain Black's tent early the next morning, reached inside the mosquito net and gently shook the moist shoulder blade he found there until Captain Black opened his eyes.

"What are you waking me up for?" whimpered Captain Black.

"They captured Bologna, sir," said Corporal Kolodny. "I thought you'd want to know. Is the mission cancelled?"

Captain Black ... peered without emotion at the map. Sure enough, they had captured Bologna.

Joseph Heller. *Catch-22.* (1961)

# more information...

Symbolization occupies large sections of any of the previously cited texts. A broader semiological approach to map and graphic symbols can be found in the wondrous work of Jacques Bertin, *Graphics and Graphic Information Processing* (Walter de Gruyter, 1981) and the *Semiology of Graphics* (University of Wisconsin Press, 1983). Eduard Imhof's *Cartographic Relief Presentation* (De Gruyter, 1982) is the classic text on terrain symbolization. For early examples of many of the map symbolization methods reviewed in this chapter, see Arthur Robinson's *Early Thematic Mapping in the History of Cartography* (University of Chicago Press, 1982).

Two surveys of information graphics and media maps with many colorful exemplars are Nigel Holmes, *Pictorial Maps: History, Design, Ideas, Sources* (Watson-Guptill, 1991) and Peter Wildbur's *Information Graphics: A Survey of Typographic, Diagrammatic, and Cartographic Communication* (Van Nostrand Reinhold, 1989).

A terrific collection of maps by artists – revealing how they have appropriated map symbolization for their own purposes – can be seen in Katharine Harmon's *You Are Here: Personal Geographies and Other Maps of the Imagination* (Princeton Architectural Press, 2004).

**Sources:** AIDs data (p. 200) from Abraham Verghese's *My Own Country: A Doctor's Story* (Vintage, New York, 1994) and Peter Gould's *The Slow Plague: A Geography of the AIDs Pandemic* (Blackwell, 1993). Hate group data (pp. 202-203) from the Southern Poverty Law Center (www.splcenter.org). Iraq war casualties (pp. 204-205) from Iraq Coalition Casualty Count (icasualties.org). Hate crimes data (p. 206) are from the U.S. Federal Bureau of Investigation (www.fbi.gov). Maps of Pennsylvania AIDs data (pp. 210-211) are based on maps in MacEachren's *Some Truth With Maps* (Association of American Geographers, 1994). Suicide rate cartogram (p. 218) and multivariate cartogram (p. 219) redrawn from Dent's *Cartography: Thematic Map Design* (1998). The "uni-value" cartogram, designed by Catherine Reeves, is redrawn from *Globehead! The Journal of Extreme Geography* (1994). Contiguous and non-contiguous cartograms (p. 219) are redrawn from Judith Tyner's *Introduction to Thematic Cartography* (Prentice Hall, 1992).

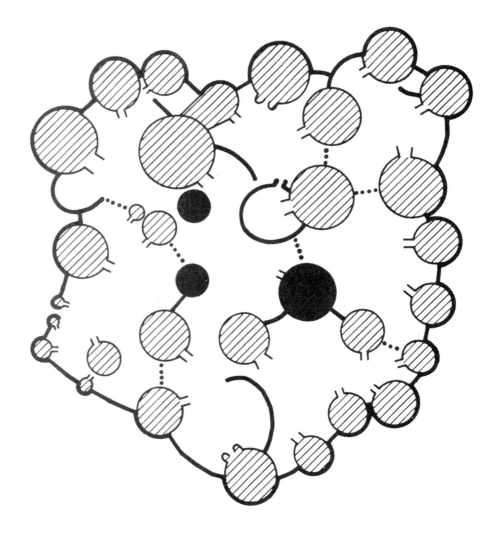

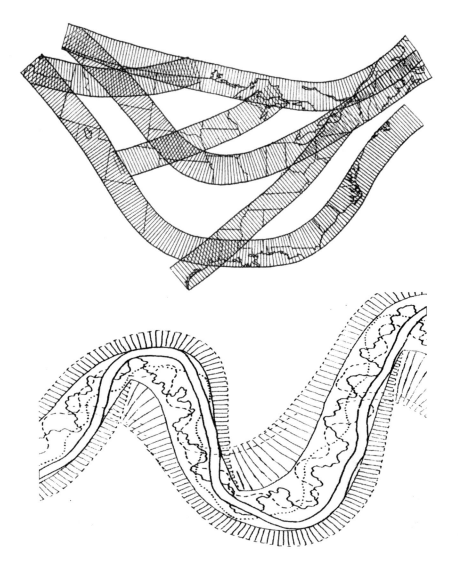

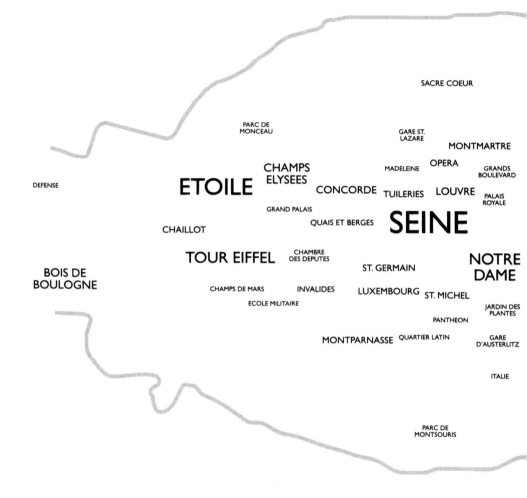

SACRE COEUR

PARC DE
MONCEAU

GARE ST.
LAZARE

MONTMARTRE

CHAMPS
ELYSEES

MADELEINE

OPERA

GRANDS
BOULEVARD

DEFENSE

ETOILE

CONCORDE

TUILERIES

LOUVRE

PALAIS
ROYALE

GRAND PALAIS

QUAIS ET BERGES

SEINE

CHAILLOT

TOUR EIFFEL

CHAMBRE
DES DEPUTES

ST. GERMAIN

NOTRE
DAME

BOIS DE
BOULOGNE

CHAMPS DE MARS

INVALIDES

LUXEMBOURG

ST. MICHEL

ECOLE MILITAIRE

JARDIN DES
PLANTES

PANTHEON

MONTPARNASSE

QUARTIER LATIN

GARE
D'AUSTERLITZ

ITALIE

PARC DE
MONTSOURIS

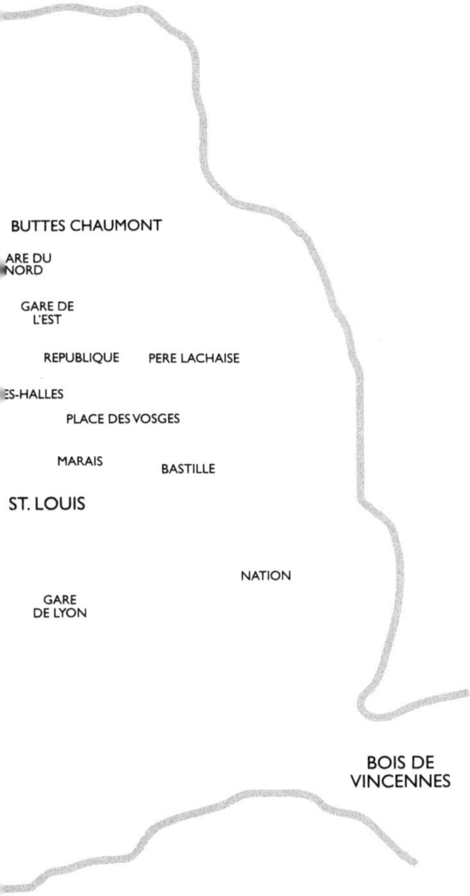

BUTTES CHAUMONT

ARE DU
NORD

GARE DE
L'EST

REPUBLIQUE     PERE LACHAISE

ES-HALLES

PLACE DES VOSGES

MARAIS

BASTILLE

ST. LOUIS

NATION

GARE
DE LYON

BOIS DE
VINCENNES

*How can the words mean*
*more than they say?*

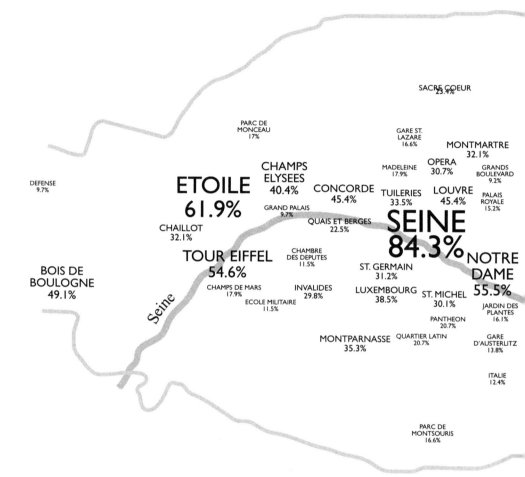

SACRE COEUR
23.4%

PARC DE
MONCEAU
17%

GARE ST.
LAZARE
16.6%

MONTMARTRE
32.1%

MADELEINE
17.9%

OPERA
30.7%

GRANDS
BOULEVARD
9.2%

DEFENSE
9.7%

ETOILE
61.9%

CHAMPS
ELYSEES
40.4%

CONCORDE
45.4%

TUILERIES
33.5%

LOUVRE
45.4%

PALAIS
ROYALE
15.2%

GRAND PALAIS
9.7%

CHAILLOT
32.1%

QUAIS ET BERGES
22.5%

SEINE
84.3%

TOUR EIFFEL
54.6%

CHAMBRE
DES DEPUTES
11.5%

NOTRE
DAME
55.5%

ST. GERMAIN
31.2%

BOIS DE
BOULOGNE
49.1%

CHAMPS DE MARS
17.9%

INVALIDES
29.8%

LUXEMBOURG
38.5%

ST. MICHEL
30.1%

ECOLE MILITAIRE
11.5%

JARDIN DES
PLANTES
16.1%

PANTHEON
20.7%

Seine

MONTPARNASSE
35.3%

QUARTIER LATIN
20.7%

GARE
D'AUSTERLITZ
13.8%

ITALIE
12.4%

PARC DE
MONTSOURIS
16.6%

BUTTES CHAUMONT
24.4%

ARE DU
NORD
14.7%

GARE DE
L'EST
15.6%

REPUBLIQUE          PERE LACHAISE
14.3%                12.9%

S-HALLES
10.1%      PLACE DES VOSGES
             18.4%

MARAIS
26.2%            BASTILLE
                 22.1%

T. LOUIS
31.7%

NATION
12%

GARE
DE LYON
18.4%

Seine

BOIS DE
VINCENNES
38.1%

10

# Type on Maps

***Type on a map means what it says, and also means what it
shows.*** A word alone means something, but its size – big or small –
on this map also means something. Stanley Milgram was a psychologist
interested, among other things, in the mental images people made of
their environment. In one study he asked over 200 Parisians to draw
maps of Paris. This map shows the 50 most cited elements. The name
of each locale is shown in a size proportional to the number of subjects
who included it in their hand-drawn maps of Paris. In this case, type
makes the point all by itself.

233

# Type on Maps

Type is subtle but important, shaping the look and effectiveness of a map. The meaning of words on maps is complemented by the *look* of the word: variations in *type style, size, weight, form,* and the *placement* of type on maps imparts additional information. The effective use of type on maps requires some understanding of the anatomy of type.

**Typeface and font:** a typeface is a collection of letters and numbers with a unique design. A font is a subset of a typeface, and includes all letters and numbers of a specific size. **Font** is often used to mean typeface. Helvetica is a different typeface (font) from Times Roman.

**Serifs:** finishing strokes added to the ends of letter: Helvetica has no serifs (sans serif font); Times Roman does (serif font).

**Typeface (font)** is Times Roman:

Making ] X-height

descender

Maps Point size =48

Serif

**Typeface (font)** is Helvetica:

Point size =48

Sans serif

ascender

Making

X-height [ Maps

**X-height** is the height of the most compact letters in a type face, such as an a, o, or e. Type with a greater X-height is typically easier to read. An **ascender** is the portion of certain letters that rises above the X-height, such as in the letters d or f. A **descender** is the portion of certain letters that falls below the X-height, such as in the letters g or p.

**Point size:** type size is measured in points, where 72 points = 1 inch. Type size is determined by the height of the original lead foundry block, and is not the same as the height of the letter: 8 point Times Roman is smaller than 8 point Georgia.

234

The use of type on a map has **two** facets:

**1**

**What the type looks like:** type as a graphic symbol

• type variables:

Type Style  **Type Weight**

**Type size**  *Type Form*

• use graphic variations in type to communicate something!

●**Larger City**

●Smaller City

**Type on maps can be used to differentiate...**

**Qualitative** data
   ex) types of trees
   ex) religious denominations

   ✓ type **style.**
   ✓ type **form** (spacing, italics,
      color hue).

**Quantitative** data:
   ex) size of population
   ex) amount of traffic

   ✓ type **size.**
   ✓ type **weight.**
   ✓ type **form** (case, color value).

**2**

**Type placement:** the arrangement of type on a map

• carefully consider how type is positioned in terms of the
   object or symbol it refers to:

Rome**o**

Romeo? **No!** Rome:

   ○Rome

# **1** What the type looks like:
## type variables

# type style

**Type Style** variations affect the overall "look" of the map:

> **Type style** is best used to symbolize **nominal (qualitative)** information:
> - ✓ **serif** type (Times Roman) used for the historical information on a historical map.
> - ✓ **sans serif** type (Gill) used for natural features on a historical map.

**Poor use of type style:**

**Good use of type style:**

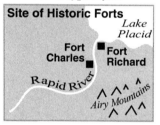

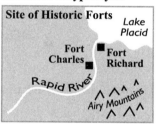

> **Type style** may also be used to symbolize the **overall "look"** of the map:
>
> - ✓ **serif** type (Times Roman) implies tradition, dignity, and solidity.
> - ✓ **sans serif** type (Gill) implies newness, precision, authority.

Considerations when using **type style variations**:

> - ✓ it is not good to combine more than two type styles on a map.
> - ✓ evaluate compatibility if serif and sans serif type styles are combined.
> - ✓ avoid combining two serif or two sans serif type styles on one map.
> - ✓ serif type is easier to read in blocks of text.
> - ✓ *avoid decorative type styles as they are very difficult to read.*

**Poor use of type style:**

**Good use of type style:**

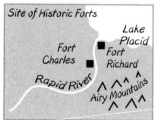

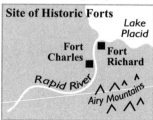

236

# *type size*

**Type Size** variations imply **ordered** (quantitative) differences:

**Larger** sizes imply more importance or greater quantity:

## Columbus

**Smaller** sizes imply less importance or less quantity:

East Beeler

**Poor use of type size:**   **Good use of type size:**

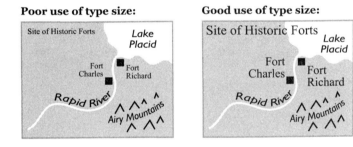

Considerations when using **type size variations**:

- ✓ type **less than 6 points** in size will be difficult to read.
- ✓ type **larger than 24 points** is getting a bit large for most page-size maps.
- ✓ use a **2 point difference** in small type sizes and **3 point difference** in medium and larger type sizes if you want the difference noticed.
- ✓ most people have difficulty distinguishing any more than **five categories** of data symbolized by type size on a map.

# type weight

**Type weight** variations imply **ordered** (quantitative) differences.

> **Bold** type implies a greater importance or quantity:
> > ✓ **bold** for county names; regular weight for township names on state map.

> Regular type implies a lesser importance or quantity:
> > ✓ population categories: 8 pt. regular, **8 pt. bold.**

**Poor use of type weight:**          **Good use of type weight:**

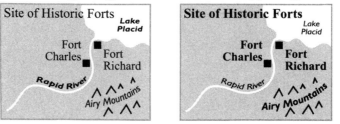

Considerations when using **type weight variations**:

> ✓ **bold** implies **more**; regular implies **less.**
> ✓ **bold** implies **power** and **significance**, yet some bold type looks **pudgy.**
> ✓ **bold** type helps legibility if type is grey.
> ✓ if you feel the urge to **<u>underline</u>** type - **STOP** - use **bold** instead.

# *type form*

**Type form** variations can imply both **nominal** (qualitative) and **ordered** (quantitative) differences.

**Spacing:** condensed vs. e x t e n d e d:
   ✓ **nominal:** extended type is often appropriate for naming **a r e a** features.

*Italics* **vs. plain** (roman) type forms:
   ✓ **nominal:** *water features in italics.*

Poor use of type spacing/italics:    Good use of type spacing/italics:

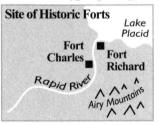
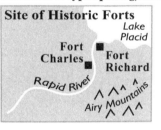

**Case:** UPPER CASE vs. lower case:
   ✓ **nominal:** MOUNTAIN RANGES upper case, other natural features lower case.
   ✓ **ordered:** STATE NAMES in upper case, county names in lower case.

Poor use of type case:    Good use of type case:

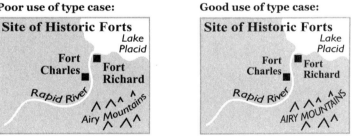

# *type form* (continued)

**Color:** color value and color hue:
- ✓ **nominal:** park names green, road names red, water names blue.
- ✓ **ordered:** background information grey, foreground information black.

Poor use of type value:             Good use of type value:

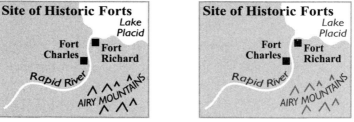

Considerations when using **type form variations**:

- ✓ **UPPER CASE** type implies **more** but is **harder to read** than lower case.
- ✓ the use of grey type must be carefully considered: small grey type is **difficult to see and read.**
- ✓ use white or light grey type (reversed type) if type is to be legible over a **dark background.**
- ✓ **condensed** type may look too squashed together to read at small sizes.
- ✓ convention suggests that *italics* be used for **natural** features and **regular** type be used for **cultural** features.

SACRE COEUR

BUTTES CHAUMONT

PARC DE
MONCEAU

GARE ST.
LAZARE

GARE DU
NORD

MONTMARTRE

OPERA

GARE DE
L'EST

CHAMPS
ELYSEES

MADELEINE

GRANDS
BOULEVARD

ETOILE

CONCORDE

TUILERIES

LOUVRE

REPUBLIQUE

PERE LACHAISE

DEFENSE

PALAIS
ROYALE

GRAND PALAIS

LES-HALLES

QUAIS ET BERGES

SEINE

PLACE DES VOSGES

CHAILLOT

MARAIS

BASTILLE

TOUR EIFFEL

CHAMPRE
DES DEPUTES

NOTRE
DAME

ST. LOUIS

ST. GERMAIN

BOIS DE
BOULOGNE

CHAMPS DE MARS

INVALIDES

LUXEMBOURG

ST. MICHEL

NATION

ECOLE MILITAIRE

JARDIN DES
PLANTES

GARE
DE LYON

PANTHEON

MONTPARNASSE

QUARTIER LATIN

GARE
D'AUSTERLITZ

ITALIE

BOIS DE
VINCENNES

PARC DE
MONTSOURIS

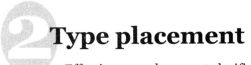 **Type placement**

Effective type placement clarifies the relationship between a label and the symbol (point, line, area) to which it refers. GIS applications often offer automated feature labeling, which normally follows traditional type placement rules.

## *labeling points*

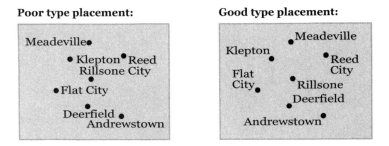

**Poor type placement:**

Meadeville•

• Klepton •Reed
Rillsone City
•
•Flat City
•
Deerfield •
Andrewstown

**Good type placement:**

•Meadeville
Klepton
•
Flat •Reed
City City
•
Flat
City• •Rillsone
•Deerfield
•
Andrewstown•

When labeling point symbols on a map, start at the **center** of the map and work outward. For each symbol, follow these **priorities**: 1 = best, 8 = worst:

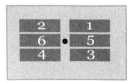

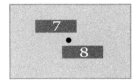

**Poor type placement:**

**Good type placement:**

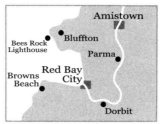

**Poor type placement:**

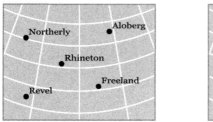

**Good type placement:**

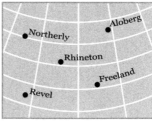

Type placement can reflect characteristics of the labeled location:

    ✓ label ports and harbor towns on the sea.
    ✓ label inland towns on the land.
    ✓ label land features on land; water features on water.
    ✓ label towns on the side of river they are located.
    ✓ align type to graticule (latitude line) if graticule is included.

# labeling lines

When labeling lines, **curve** or **slant** the type to follow the symbol. Focus on three goals: **above, horizontal, repeat**.

**Poor type placement:**

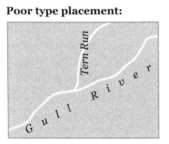

**Good type placement:**

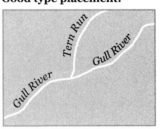

As there are fewer type descenders than ascenders, place labels on lines **above** the symbol and fit descenders into the symbol.

**Poor type placement:**

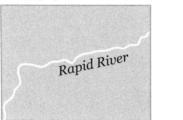

**Good type placement:**

**Horizontal** type is easiest to locate and read. Never place type upside down. If vertical, place first letter of label at the bottom.

**Poor type placement:**

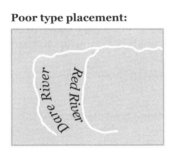

**Good type placement:**

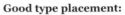

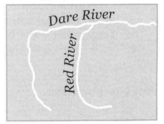

**Repeat**, rather than space out the label along a linear symbol.

**Poor type placement:**

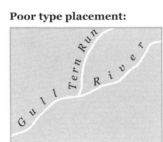

**Good type placement:**

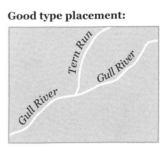

# *labeling areas*

When labeling areas on maps, **curve** and **space** the type to fit the areas to ensure that the area and the label are clearly associated.

- ✓ entire area label should follow a gentle and smooth curve.
- ✓ keep area labels as horizontal as possible (easier to read).
- ✓ avoid vertical and upside-down labels (harder to read).
- ✓ keep labels away from area borders.
- ✓ avoid hyphenating or breaking up area labels.

**Poor type placement:**  **Good type placement:**

**Overlapping** areas can be distinguished by varying type variables such as size, weight, and form.

Poor type placement:                    Good type placement:

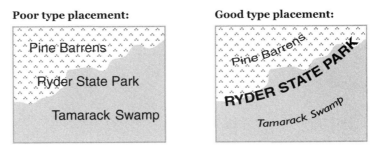

**Linear** areas should be labeled like line symbols.

Poor type placement:                    Good type placement:

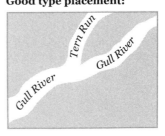

He was certain that in the pattern of its lines and letters this map contained the answer to the old conundrum of his life in Raintree County. It was all warm and glowing with the secret he had sought for half a century. The words inscribed on the deep paper were dawnwords, each one disclosing the origin and essence of the thing named. But as he sought to read them, they dissolved into the substance of the map.

Ross Lockridge, Jr. *Raintree County.* (1948)


"I can't read the names on this [map] because they are all in English."

Christian realized he would have to show his friend how to read a map. "The top is north," he said. "The little circles are town and villages. Blue means rivers and lakes. The thin lines are roads and the thick ones railways."

"There's nothing at all here," said Big Tiger, pointing to one of the many white patches.

"That means it's just desert," Christian explained. "You have to go into the desert to know what it looks like."

Fritz Muhlenweg. *Big Tiger and Christian.* (1952)


An Irish railroad gang, coming on a lake in Arkansas that the French had named L'Eau Froid, struggled awhile, and left it on the map as Low Freight.

Alistair Cooke. *Alistair Cooke's America.* (1973)


248

# more information...

Most cartography texts devote some space to type on maps. Again, Borden Dent's type chapter in *Cartography: Thematic Map Design* (1998) is the best.

The idea for the good/poor type placement pairs is from a 1975 article by the famed Swiss cartographer Eduard Imhof: "Positioning Names on Maps" (*The American Cartographer,* 2:2, pp. 128-44). That good/poor pairs idea sorta leaked into the rest of this book: thanks, Ed!

To learn more about type, see Warren Chappell's *A Short History of the Printed Word* (Knopf, 1970). Robert Binghurst's *Elements of Typographic Style* (Hartley and Marks, 2004) is a classic overview of type. Edward Catich's *The Origin of the Serif* (2nd ed., St. Ambrose University Press, 1991) is one of the most thought-provoking books about letters we know of. The issue of *Visible Language* (Autumn 1982) with Douglas Hofstader and other type designers' responses to Donald Knuth's article about a "meta-font" is interesting also.

**Sources:** The type map of Paris is based on one from a chapter Stanley Milgram published in H. Proshansky et al., *Environmental Psychology*, 2nd ed. (Holt, Reinhart, and Winston, 1976). The type placement pairs are based on the Imhof article cited above.

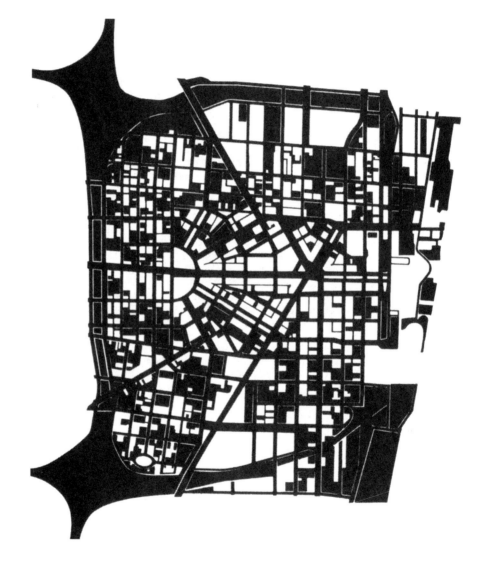

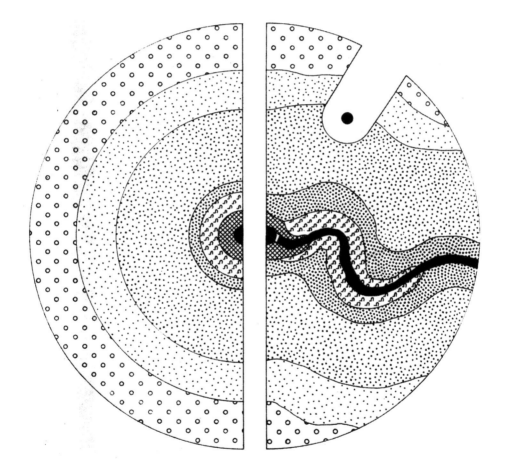

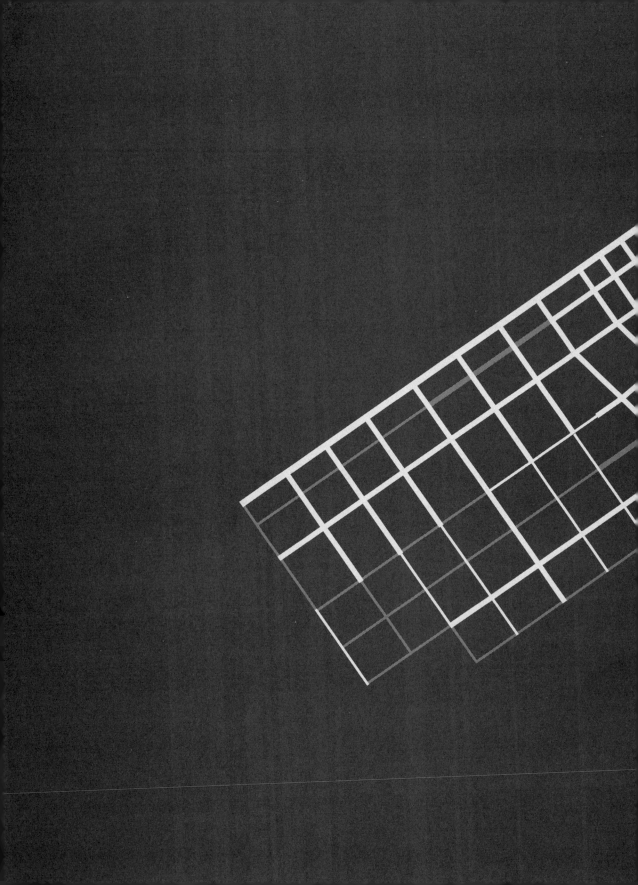

Do you need to move beyond
*black and white?*

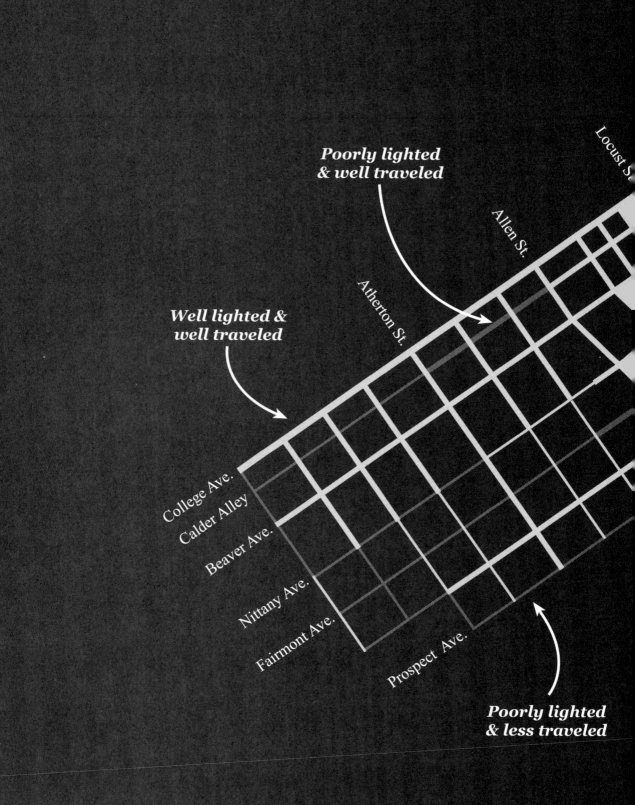

Grant St.

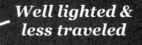

**Well lighted &**
**less traveled**

# 11

# Color on Maps

*You don't need color to make excellent maps.* This "night" map of State College, Pennsylvania, shows how well streets are lighted at night and how many people are around. The map helps people choose a safe route, and the map *wouldn't do that any better if color were used.* Color can be a great addition to a map, but is often *not* necessary.

Well lighted & well traveled
Well lighted & less traveled
Poorly lighted & well traveled
Poorly lighted & less traveled

255

# Color on Maps

Color is a **vital** and **vexing** part of making maps. Prior to the computer, making color maps was difficult and expensive. With computers, color is always an option and is often used poorly and even when it is not necessary. Yes, you can easily use color on you map, but ask yourself: *Is it really necessary?* If so, then at least use color well.

*no!*
*no!*
*no!*

## Election 2004

The *fruity* colors on the above map may appeal to those with dubious tastes, but they make the data *tough to understand:*

✓ which counties have the **highest** rates?
✓ which counties have the **lowest** rates?

Switch to *red & blue* and ask the same questions of the map. The reader has a much *easier* time interpreting the data!

**Large** Bush Win
**Medium** Bush Win
Small **Bush Win**
Small **Kerry Win**
**Medium** Kerry Win
**Large** Kerry Win

*ya!*
*ya!*
*ya!*

# *"Color is a cartographic quagmire!"*

Mark Monmonier, How to Lie with Maps, 1996

Color is a "cartographic quagmire" because it is often misused – especially since color has become ubiquitous with computer map making.  Color is also a problem because:

- ✓ **color terminology is confusing, with no single standard.**
- ✓ **there are many ways to define and specify colors.**
- ✓ **you can make very effective maps using black and greys.**

## Extract yourself from the **quagmire**:

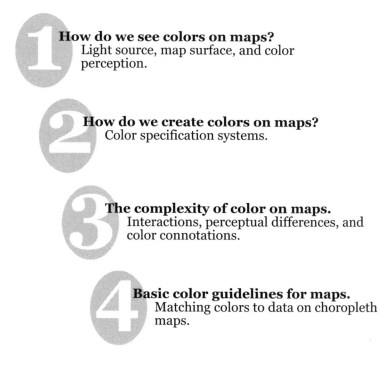

**How do we see colors on maps?**
Light source, map surface, and color perception.

**How do we create colors on maps?**
Color specification systems.

**The complexity of color on maps.**
Interactions, perceptual differences, and color connotations.

**Basic color guidelines for maps.**
Matching colors to data on choropleth maps.

# 1 How do we see colors on maps?

If we understand how people see colors on maps, we can use that understanding to guide the use of colors when making maps.

## *light source*

The colors on a map vary as the **light source** varies. The *same* colors will look *different* when viewing a map

- ✓ under *incandescent* versus *florescent* lighting.
- ✓ as the *intensity* of the light source varies.

When selecting colors for a map, consider the conditions under which your map will be viewed:

- ✓ *low-intensity lighting:* use more intense, saturated colors.
- ✓ *high-intensity lighting:* use less intense, less saturated colors.
- ✓ look *critically* at your map under lighting conditions similar to those of your map's audience and adjust the colors to suit.

## *map surface*

The colors on a map vary as the **surface** the map is displayed on varies. The *same* colors will look *different* when viewing a map

- ✓ on *glossy* versus *matte* surfaced paper.
- ✓ on *paper* versus *projected* versus on a *computer monitor.*

When selecting colors for a map, consider the effect of the map surface:

- ✓ *glossy paper:* will make colors more intense and vibrant.
- ✓ *matte paper:* will make colors less intense and dulled.
- ✓ *projected:* depending on the intensity of the projector, the colors may be significantly more or less intense than you expect.
- ✓ *computer monitor:* will make colors intense and vibrant, as the color on computer monitors is projected rather than reflected (which is the case with paper maps).
- ✓ look *critically* at your map on the surface the map is intended to be presented on and adjust the colors if necessary.

# seeing color

Pigment cells in our eyes are sensitive to blue, green, and red wavelengths with overlap, so we can sense the entire spectrum (red, orange, yellow, green, blue, indigo, violet). One way to understand how people perceive colors is in terms of three **dimensions of color perception:** hue, value (lightness), and intensity (saturation, chroma).

## color hue hue hue

A **hue** is the name for our human experience of particular electromagnetic energy wavelengths.

> ✓ **blue, green, red, yellow, orange, violet**

Different hues are *qualitatively* different: red is not more or less than green. ***Hue is appropriate for showing qualitatively differing map elements, such as a river and a road.***

## color value value

**Value** is the perceived lightness or darkness of a hue.

> ✓ light red, red, **dark red**

Different values are *quantitatively* different: light red is less than dark red. ***Value is appropriate for showing quantitatively differing map elements such as elevations or different-sized roads.***

## color intensity intensity

**Intensity** (saturation) describes the purity of a hue.

> ✓ **high intensity, intensity, low intensity**

Intensity is very subtle. ***Intensity is appropriate for binary data - is a country a member of NATO (high-intensity color) or it is not (low-intensity color)?***

# ②How do we create colors on maps?

Further complicating things is the fact that the specification and production of colors is often very different from the way we see them.

## Color Specification Systems

Color specification systems are schemes that *organize* and help *produce* different colors. There are many different color specification systems, and map makers will encounter all of the major systems. Three major categories of color specification systems are important to understand:

✓ predefined color systems.
✓ perceptual color systems.
✓ process color systems.

## *predefined systems*

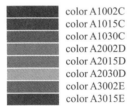

color A1002C
color A1015C
color A1030C
color A2002D
color A2015D
color A2030D
color A3002E
color A3015E

**Predefined color specification systems** are like the paint chips you find at a paint store. Thousands of predefined colors are specified by names or code numbers. Predefined colors are used if you have a commercial printer who asks for such colors (often referred to as "spot" color). Predefined colors will be converted to another color specification system if you are printing on a computer printer, or using a commercial printer who uses process color printing.

## *perceptual systems*

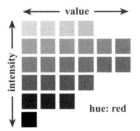

value

intensity

hue: red

**Perceptual color specification systems, such as Munsell,** are based on human perceptual abilities. Perceptual tests have resulted in a select set of colors that the average person can differentiate: in other words, no two colors in the Munsell system look exactly alike. The Munsell system consists of a series of color samples, each a single hue with varying value and intensity. The Munsell system is excellent for selecting appropriate colors, but will be converted to another system in order to print.

260

# process systems
## commercial and computer printing

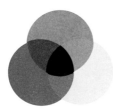

**Process color specification systems** use three or four colors to create all other colors. When you combine cyan (C), magenta (M), and yellow (Y) you produce black - all light is absorbed or subtracted from your vision. Black is often added as a fourth color (K, thus CMYK). These *subtractive primaries* are used by commercial printers, and are common on inkjet computer printers. Different percentages of CMY and K produce thousands of other colors. The CMYK color system should be used for most commercially printed maps.

# process systems
## computer monitors

**Computer monitors** also use three colors to create all other colors. When you combine red (R), green (G), and blue (B) you produce white - all light is reflected. These *additive primaries* use different amounts of R, G, and B to produce thousands or millions of other colors. The hexidecimal color specifications used in HTML are really RGB: the first two digits are red, second two digits green, and third two digits blue. 00 is no color and FF is maximum color. The RGB color system should be used for maps printed with computer printers. RGB will have to be converted into CMYK or predefined color if you plan to print with a commercial printer.

# ③ The complexity of color use on maps

The use of colors on maps is complex: colors interact with surrounding colors, there are perceptual differences among map viewers, and color has symbolic connotations.

## *color interacts with surrounding colors*

### Simultaneous Contrast

The appearance of any color on a map depends on the colors that surround it. This optical illusion makes the grey dot on the top look slightly darker than the grey dot below (for most people).

If the background of a map has varying colors, check that the symbols that are supposed to be the same color look the same everywhere on the map.

### Purity of Hues

When used together on a map, some hues look pure, while other hues look like mixtures. Green and red seem to be relatively pure compared to orange or purple, which seem to be a mix.

Consider the purity of hues when combining colors on a map. If your goal for your map is to imply distinctive differences, use pure hues (green, red, blue). If your goal is to imply less distinctive differences, used mixed hues (orange, brown).

**Poor use of purity of hues:**

## 2002 Township Elections
Reed County, WI

● Republican Win
● Democrat Win

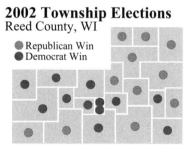

**Good use of purity of hues:**

## 2002 Township Elections
Reed County, WI

● Republican Win
● Democrat Win

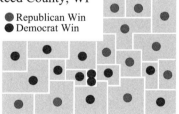

262

## Visual Difference

It is slightly easier to see the colors of map elements when the map background is a pure, monochromatic hue (green, red) rather than a less pure hue (brown, purple). If you intend for your map to clearly distinguish some data from other data, use colors that have high visual difference. Less visual difference is also useful if your goal is to suggest less difference between data.

**Poor visual difference:**

### 2002 Township Elections
Reed County, WI

- Republican Win
- Democrat Win

**Good visual difference:**

### 2002 Township Elections
Reed County, WI

- Republican Win
- Democrat Win

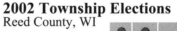
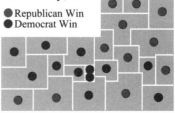

# perceptual differences among individuals

The appearance of color on a map varies based on **who** is looking at it:

### Older map viewers

- ✓ have difficulty seeing colors and need more saturated colors.
- ✓ have particular difficulty seeing blue colors.
- ✓ benefit from increasing the type size a bit.

### Color-blind map viewers: for the color-blind, red and green often look the same.

- ✓ in the U.S., 3% of females and 8% of males are color-blind.
- ✓ if reds and greens show important differences on your map, be aware that a significant number of viewers *will not see these differences.*
- ✓ consider using reds and blues or greens and blues instead.

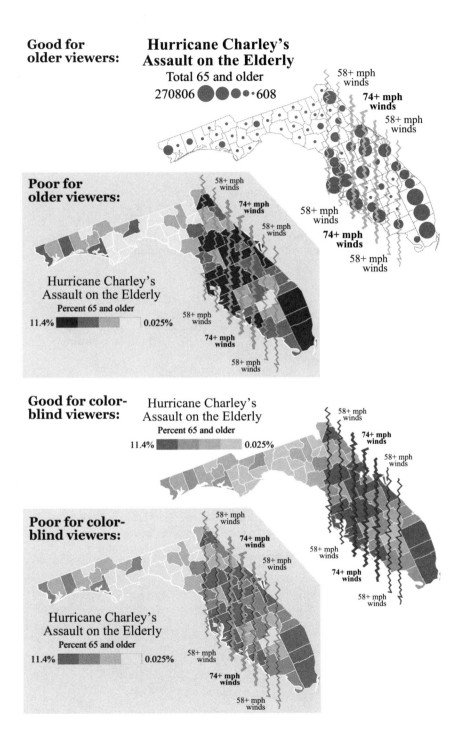

**Good for older viewers:**

**Hurricane Charley's Assault on the Elderly**

Total 65 and older

270806 ●●●●·608

58+ mph winds

**74+ mph winds**

58+ mph winds

**Poor for older viewers:**

58+ mph winds

**74+ mph winds**

58+ mph winds

Hurricane Charley's Assault on the Elderly

Percent 65 and older

11.4%          0.025%

58+ mph winds

**74+ mph winds**

58+ mph winds

**Good for color-blind viewers:**

Hurricane Charley's Assault on the Elderly

Percent 65 and older

11.4%          0.025%

58+ mph winds

**74+ mph winds**

58+ mph winds

**Poor for color-blind viewers:**

58+ mph winds

**74+ mph winds**

58+ mph winds

58+ mph winds

**74+ mph winds**

58+ mph winds

Hurricane Charley's Assault on the Elderly

Percent 65 and older

11.4%          0.025%

58+ mph winds

**74+ mph winds**

58+ mph winds

265

# color connotations

Color has *symbolic connotations*. Such connotations subtly shape viewer reactions, and should be guided by your goals for your map.

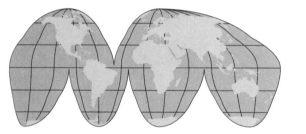

**Poor color conventions**

## *Western culture and color connotations*:

- ✓ **blue:** water, cool, positive numbers, serenity, purity, depth.
- ✓ **green:** vegetation, lowlands, forests, youth, spring, nature, peace.
- ✓ **red:** warm, important, negative numbers, action, anger, danger, power, warning.
- ✓ **yellow/tan:** dryness, lack of vegetation, intermediate elevation, heat.
- ✓ **orange:** harvest, fall, abundance, fire, attention, action, warning.
- ✓ **brown:** landforms (mountains, hills), contours, earthy.
- ✓ **purple:** dignity, royalty, sorrow, despair, richness, elegant.
- ✓ **white: purity, clean, faith, illness.**
- ✓ **black:** mystery, strength, heaviness.
- ✓ **greys:** quiet, reserved, sophisticated, controlled.

**Good color conventions**

***Cultural differences and color connotations***: the meaning
we get from different colors varies from culture to culture:

- ✓ **pink** is associated with the *feminine* in the United States and India,
  but is not in Japan.
- ✓ **green** is associated with fertility and paganism in northern Europe,
  is a sacred color for Muslims.
- ✓ **white** is associated with widowhood and unhappiness in India, and
  with mourning in China.
- ✓ **purple** is *dangerous*: Catholic Europeans see purple as a symbol of
  *death* and *crucifixion;* in some Middle Eastern cultures purple
  represents *prostitution*. Purple is also symbolic of *mysticism* and
  *spiritual* beliefs that counter Christian, Jewish, and Muslim religions.
- ✓ **blue** is one of the *safest* cross-cultural colors, possibly because it is
  the color of the sky, which stands over all peoples, and is often the
  realm of deities - heaven.
- ✓ color has **gender** associations: men prefer **blue** to **red,** women
  **red** to **blue;** men prefer **orange** to **yellow,** women **yellow** to
  **orange.**

Good in U.S.: Iraq is Dangerous:

Bad in Iraq: Iraq as Prostitute:

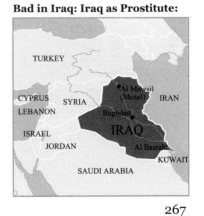

267

# Basic color guidelines for maps

Color differences should suggest differences in your data. Qualitative, binary, and ordered (quantitative) differences can be matched to colors that suggest similar variations.

**Mapping Qualitative Data**

**Poor qualitative colors (value):**

Favorite Hotdog Condiment
Majority Opinion, Oregon, 2003

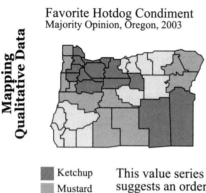

■ Ketchup
■ Mustard
□ Relish

This value series suggests an order in the data that does not exist.

**Good qualitative colors (hue):**

Favorite Hotdog Condiment
Majority Opinion, Oregon, 2003

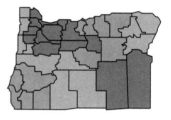

■ Ketchup
□ Mustard
■ Relish

Three hues suggest no order and reflect actual condiment colors.

**Mapping Binary Data**

**OK binary colors (value):**

Elvis Is Dead?
Majority Opinion, Oregon, 2003

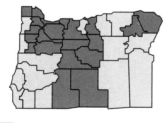

■ Yes
□ No

This pair of values suggests that **Yes** opinions are more important than **No**.

**OK binary colors (hue):**

Elvis Is Dead?
Majority Opinion, Oregon, 2003

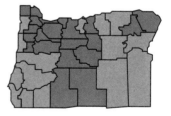

■ Yes
■ No

Two hues suggest either opinion is as important.

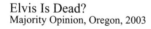

268

**Poor ordered colors (hue):**

Fallen, Cannot Get Up
per 1000 population, Oregon, 2002

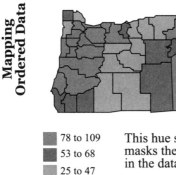

78 to 109
53 to 68
25 to 47
2 to 20

This hue series masks the order in the data.

**Good ordered colors (value):**

Fallen, Cannot Get Up
per 1000 population, Oregon, 2002

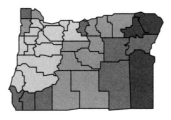

78 to 109
53 to 68
25 to 47
2 to 20

This value series reveals the order in the data.

**Mapping Ordered Data**

**OK ordered colors (value):**

Beanie Baby Sales
Percent Change, Oregon 2000-2002

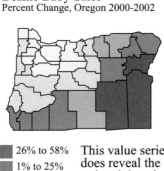

26% to 58%
1% to 25%
0 to -99%
-100% to - 499%
-500% to -1000%

This value series does reveal the ordered data, but...

**Better ordered colors (value):**

Beanie Baby Sales
Percent Change, Oregon 2000-2002

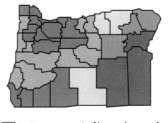

26% to 58%
1% to 25%
0 to -99%
-100% to - 499%
-500% to -1000%

A diverging value series best reveals the diverging data.

**Mapping Diverging Ordered Data**

269

The earth is a place on which England is found,
And you find it however you twirl the globe round;
For the spots are all red and the rest is all grey,
and that is the meaning of Empire Day.

G.K. Chesterton. *Songs of Education.* (1927)


I do not advance that the face of our country would change if the maps which Philadelphia sends forth all over the Union were more decently colored, but certainly it would indicate that the Graces were more frequently at home on the banks of our lovely rivers, if the engravers were able to sell their maps less boisterously painted and not as they are now, each county of each state in flaming red, bright yellow, or a flagrant orange dye, arrayed like the cover produced by the united efforts of a quilting match. When I once complained of this barbarous offensive coloring of maps, the geographer assured me that he would not sell them unless bedaubed in this way; "for, said he, the greatest number of the large maps are not sold for any purpose of utility, but to ornament the walls of barrooms. My agents write continually to me to color high." This reason was given me by one of the first geographers of the United States, who has himself a perfectly correct idea of the tasteful coloring of maps.

Francis Lieber. "On Hipponomastics: A Letter to Pierce M. Butler," *Southern Literary Messenger* 3:5. (1837)  Thanks to Penny Richards for this quote.

# *more information...*

Cindy Brewer is an expert on color for maps. A nice summary of some of her work is "Color Use Guidelines for Mapping and Visualization" in Alan MacEachren and D.R. Fraser Taylor's *Visualization in Modern Cartography* (Pergamon, 1994). Much of Brewer's work can be found in the online resource at www.colorbrewer.org.

A great article on natural color maps is Tom Patterson and Nathaniel Vaughn Kelso's "Hal Shelton Revisited: Designing and Producing Natural-Color Maps with Satellite Land Cover Data" (*Cartographic Perspectives* 47, Winter 2004, pp. 28-55, plus a portfolio of plates, pp. 69-77).

Edward Tufte has a whole chapter on "Color and Information" in his *Envisioning Information* (Graphics Press, 1990).

**Sources:** The State College night map (pp. 252-255) is based on the original created in the Deasy GeoGraphics Lab (now the Gould Center) at Penn State. The idea for the old folks hurricane maps (p. 265) came from a map published in the *News & Observer* (Raleigh-Durham, North Carolina) of August 19, 2004.

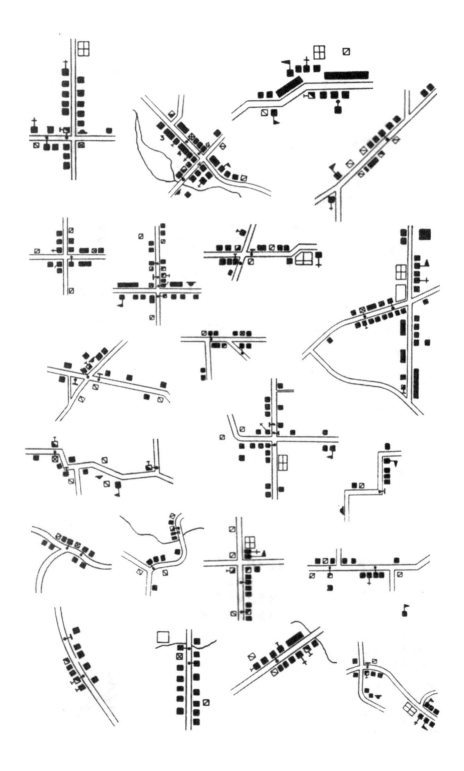

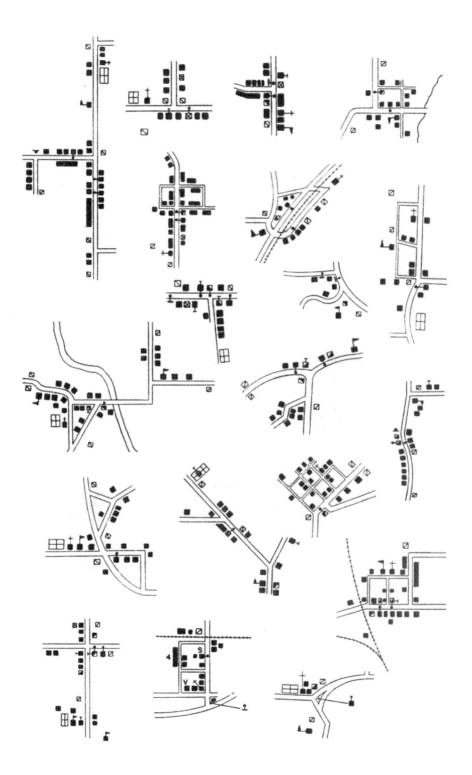

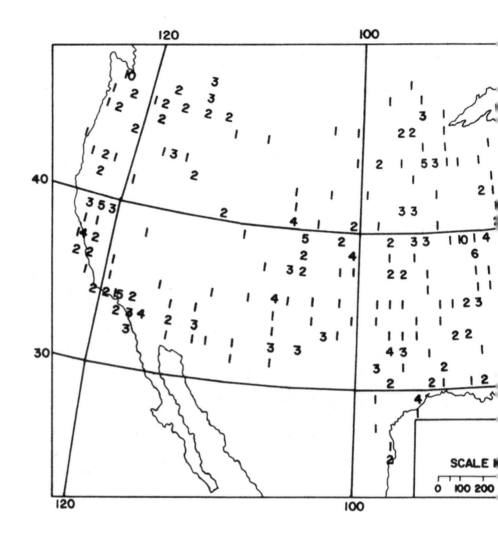

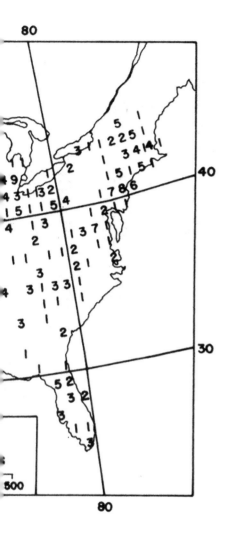

*Is it any good?*

*Who is this map for?*

*What about a title?*

*Are numbers the best way
to symbolize the data?*

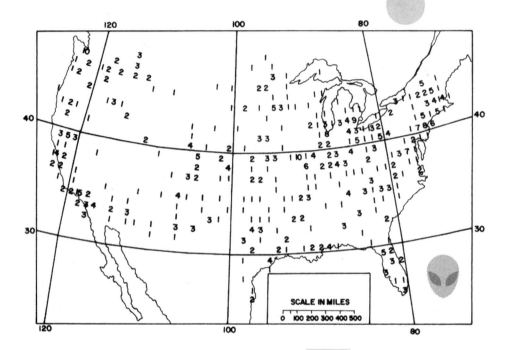

*Why are there no state
boundaries?*

*Do the most important
data visually jump out?*

*What is the projection?*

*Where are the data from?*

*Are the data at points or in areas?*

*Does the scale indicator need to be so big - and does it need to be there at all?*

*What time are the data?*

*Must the graticule be so visually prominent?*

*Has any important information been left off the map?*

# Finishing Your Map

***You're not finished until you engage your mind in a thorough critique and evaluation of your map.*** Critique and evaluation begins early and continues throughout the process of making your map. The plethora of rules and regulations about making maps along with defaults in GIS and mapping software may suggest that map making can be a mindless follow-the-rules process. *Nothing could be further from the truth!* Rules are made to be **broken** - when it serves the ultimate goals of your map. Defaults in software are made to be **discarded** - they tend to make maps created by different people with different goals and intents all look the same - and this is *very* boring. ***Making maps requires engaging your mind!***

# Finishing Your Map

With your map in front of you, thoroughly interrogate it! Change the design of your map if you are not satisfied with anything. It is easy to adjust the design of maps made with computers.

Strange lights at night and sightings of ◗◖ are a neat topic for a newspaper article. A great source of **UFO data** is the books of **Charles Fort.** Fort and his followers compiled thousands of odd events in the U.S. and around the world, and often noted the location of the event. This map must appeal to a broad audience and be as fun as the data while also being informative. The map your lackey created - which we are looking at – can be critiqued and reworked to be much better.

## *critiquing the whole map*

✓ does your map do what you want it to do?

✓ is your map suitable for your intended audience? Will they be confused, bored, interested, or informed?

✓ does the map reproduce well on its final medium? Has the potential of a black-and-white or color design been reached?

✓ describe the *overall look of the map* in terms of these word pairs, then ask: is that what I want to convey?

| | |
|---|---|
| ✓ confusing or clear | ✓ interesting or boring |
| ✓ amorphous or structured | ✓ light or dark |
| ✓ fragmented or coherent | ✓ constrained or lavish |
| ✓ random or ordered | ✓ modern or traditional |
| ✓ crowded or empty | ✓ bold or timid |
| ✓ free or bounded | ✓ subtle or blatant |
| ✓ lopsided or balanced | ✓ flexible or rigid |
| ✓ neat or sloppy | ✓ hard or soft |
| ✓ crude or elegant | ✓ tentative or final |
| ✓ high or low contrast | ✓ authoritative or unauthoratative |
| ✓ complex or simple | ✓ appropriate or inappropriate |

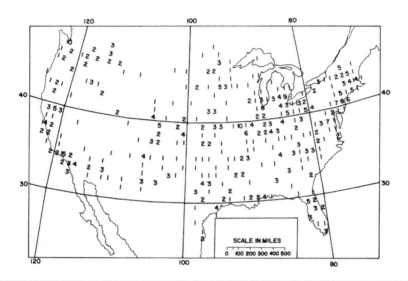

✓ only barely ... the data are there but the map is *dull* and *confusing*.

✓ the viewers of this map will certainly expect something *easier* to interpret and more *visually interesting*.

✓ the map has to be black and white, but much *more* can be done with monochrome than this pitiful map does.

✓ I don't think this is what I want to convey ...

| | |
|---|---|
| ✓ *confusing* | ✓ definitely *boring* |
| ✓ too *structured* | ✓ too *light* for *dark* phenomena |
| ✓ numbers = *fragmented* | ✓ overly *constrained* |
| ✓ numbers = *random* | ✓ blandly *traditional* |
| ✓ too *empty* | ✓ *timid* |
| ✓ over-*bounded* | ✓ too *subtle* |
| ✓ OK *balance* | ✓ dull and *rigid* |
| ✓ *neat* but dull | ✓ *hard* and edgy |
| ✓ *crude* looking | ✓ seems *tentative*, unfinished |
| ✓ contrast too *high* | ✓ *authoritative* but dull |
| ✓ *simplistic* | ✓ *inappropriate*, given map goals! |

**Undertake a systematic critique, then *redesign* the map ...**

279

# critiquing the data

- ✓ do your *data* serve your goals for the map?

- ✓ do the background data *enhance* the primary data the map is showing?

- ✓ are the data too *generalized* or too *complex,* given the map's goals ?

- ✓ does *textual explanation* or *discussion* on the map coordinate well with the map itself?

- ✓ does your map *symbolization* reflect the character of the *phenomena* or the character of the *data?*

- ✓ do map *symbols* reflect the *qualitative* or *quantitative* character of the data?

- ✓ have your data been properly *derived?*

- ✓ have you considered all aspects of the *accuracy* of your data?

- ✓ have you considered *copyright* issues?

# critiquing the framework

- ✓ what are the characteristics of your map's *projection,* and is it appropriate for the data and map goals?

- ✓ is the *scale* of the map (and inset) adequate, given your goals for the map?

- ✓ is the *coordinate system* appropriate and noted on the map?

# the critique...

✓ the data are fine but need much more contextualization.

✓ since the data are for the U.S. only, the map needs U.S. state boundaries and can drop the non-U.S. coastlines.

✓ the data could be more *complex* - possibly add some additional data that help understand the UFO data.

✓ the map lacks any *explanatory text* - thus is confusing. It would also help to know more about the data.

✓ numbers are not a horrible way to symbolize these data, but other symbols may be *better* and *more interesting*.

✓ numbers visually vary *qualitatively,* and these are *quantitative* data.

✓ the map uses *total* numbers, *generalized* to 1° by 1° cells.

✓ Charles Fort was a famed collector of such data, and the compilers and publisher of the data seem legitimate.

✓ facts can't be copyrighted, and we are re-creating the map.

# the critique...

✓ the map is equal-area (an Albers, to be precise), which is the most appropriate *property* to preserve on a map like this.

✓ the *scale* choice for the map - extent of area covered - is fine.  The scale bar is too big and probably not necessary.

✓ with this map, including *coordinate* information isn't important.  Best to leave it off!

# critiquing titles & legends

✓ is your map *title* clear and coherent and focused on what your map is about?

✓ does your legend include all symbols that are *not self-explanatory?*

✓ have you included information, if your goals for the map call for it, on *map authorship, projection used, date of map,* and *orientation?*

# critiquing map design

✓ does the *layout* of your map take into account focus, balance, and a structured grid?

✓ is the *intellectual hierarchy* of your map's information reflected in the *visual hierarchy* created on the map?

✓ is the level of *generalization,* and the *classification* of data, appropriate given your goals for your map?

✓ have map *symbols* been chosen to reflect the logical guidelines suggested by the visual variables? If symbolizing data in areas, is the most appropriate method used?

✓ does the overall look of your map's *type* match the map goals? Has the type been manipulated to convey additional information?

✓ do your *black-and-white* or *color* choices add to the ability of your map to convey information? Have they been chosen to enhance what you are trying to do with your map? Do they *reproduce* well on the final medium?

✓ have you *evaluated* and *critiqued* your map throughout the process of making it and when it is finished, and implemented changes where necessary?

# the critique...

- ✓ where is the *title?* The map has no *explanatory text* at all, and that is not a good thing.

- ✓ the numbers, while not the best way of symbolizing this data, are self-explanatory.

- ✓ more details need to be added – a *description* of the data and their *source* and *date,* and the *map projection* (in case someone wants to digitize it and put it into GIS).

# the critique...

- ✓ the layout is *very balanced* and *very boring,* and does nothing to enhance the map and its curious data.

- ✓ *map-crap* that is *not* that important - the coordinate grid and the scale bar and box - mar the map's design.

- ✓ the data are generalized to 1° by 1° cells - which is fine for the scale of this map. *Classification* may help.

- ✓ numbers are OK as symbols here - most readers understand numbers. But they are *not visually quantitative.* Graduated symbols or a choropleth map are alternatives to the number symbols.

- ✓ beyond being symbols on the map, type here is *spartan* and *underutilized.* Type that enhances the feel of the funky data would be good.

- ✓ the map is truly black and white with *no use of greys,* which would allow for additional data to be included on the map. Better use of greys will also make the map *look better* and *more sophisticated.*

- ✓ obviously, whoever made this map did not do much evaluation and critiquing while creating it. ***Time for a revision!***

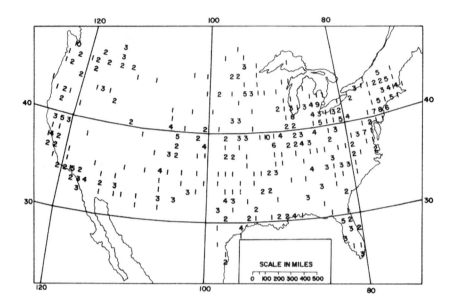

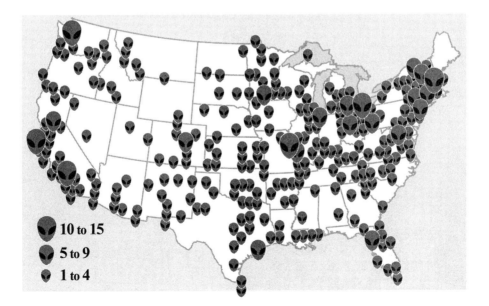

285

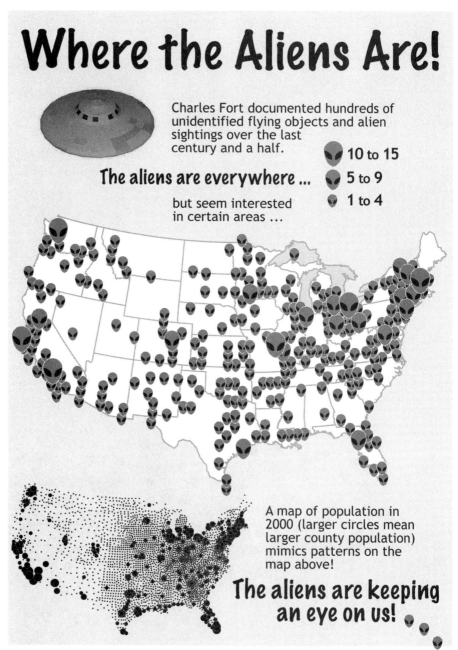

# Where the Aliens Are!

Charles Fort documented hundreds of unidentified flying objects and alien sightings over the last century and a half.

10 to 15
5 to 9
1 to 4

The aliens are everywhere ...

but seem interested in certain areas ...

A map of population in 2000 (larger circles mean larger county population) mimics patterns on the map above!

The aliens are keeping an eye on us!

Source: Charles Fort. 1974. *The Complete Books of Charles Fort.* New York: Dover Publications
Source: M. Persinger and G. Lafreniere. 1977. *Space-Time Transients.* Chicago: Nelson-Hall

is a fluid art

# Index

perceptual differences among individuals, 264–265
connotations, 266–267
color and qualitative data, 268
and binary data, 268
and ordered data, 269
and diverging ordered data, 269
Color, on paper, as final medium, 37
Columbus, Ohio, 122–125
Compilation, of data, 56
Compromise projection, 108–109
Computer monitor, as final medium, 35
Computers, making maps without, 75
Connotations, color, cultural differences, 266–267
Conrad, Joseph, 138
*Continents and Islands of Mankind*, map, 158–161
Contrast, simultaneous, 262
Cooke, Alistair, 248
Coordinates, 92–93, 112–116
matching, 113
Cosa, Juan de la, 8
Crestview Alley Debris Survey, map, 209
Cultural differences in color connotations, 266–267

# D

DALIS (Delaware [Ohio] Area Land Information System), 79
Data, 52–65. *See also* Area data, Generalization and classification, Graphing data, Line data, Point data, Qualitative data, Quantitative data
accuracy of, 63
at addresses, 57

appropriate for choropleth maps, 212
for cartograms, 218
census, 56
and color, 268–269
compilation, 56
copyright of, 65
critiquing data, 280
for dot maps, 220
and GIS, 64
using GPS, 58
for graduated symbol maps, 214
individual or aggregate, 200
layers, 55
metadata, 64
organization, 60
phenomena and, 54
primary and secondary, 59
quantifying, 61
and remote sensing, 58
for surface maps, 222
transforming, 62
Depth, in hierarchies, 146
Design guides for hierarchies, 149
Design Map, 77
Detail, in hierarchies, 150
*The Devil's Dictionary*, 44
Dickens, Charles, 188
Dimension change, in generalization, 169
Directional indicator, in layout, 129
Displacement, in generalization, 167
Distorting circles, 96–97
Diverging ordered data, color for, 269
Documentation, 41
Dot maps, 220–21
Drought, map, 33
*Dumb and Dumber*, 118
Dunne, Henry, 118

297

# Looking at maps...

A series of interesting maps is placed at the beginning of this book and between chapters. Good map makers always look at cool maps for ideas. Most of these maps have been modified to remove distracting data: please don't use them as real data. The sources of these maps are noted below, in the order in which they appear.

## *Maps before title page:*

"The Distribution of Population in Romania (1930)." J. M. Houston. 1953. *A Social Geography of Europe.* London: Duckworth.

Richard L. Miller, *Under the Cloud: The Decades of Nuclear Testing.* Two-Sixty Press, 1999.

"The Effects of Opencast Mining in New Caledonia." I. G. Simmons. 1989. *Changing the Face of the Earth.* Oxford: Blackwell. p. 237.

"Traffic and Power for a Danube Valley Authority." George Kiss. 1947. "TVA on the Danube?" *Geographical Review* 37:2. pp. 274-302.

"Interdigitation of Religions in Africa." Erwin Raisz. 1938. *General Cartography.* New York: McGraw-Hill.

"Lake Passengers Landed at Lake Michigan Ports, 1914." Ray Whitbeck. 1921. *The Geography and Economic Development of Southeastern Wisconsin.* Madison: State of Wisconsin.

"Dunes on Moenkopi Plateau." John Hack. 1941. "Dunes of the Western Navajo Country." *Geographical Review* 31:2. pp. 240-263.

"Passenger and Freight Railroad Networks, 1946." Charles Hitchcock. 1946. "Westchester-Fairfield: Proposed Site for the Permanent Seat of the United Nations." *Geographical Review* 36:3. pp. 351-397.

## *Maps preceding each chapter:*

"Pattern Tests." Michael McCullagh and John C. Davis. 1972. "Optical Analysis of Two-Dimensional Patterns." *Annals of the Association of American Geographers* 62:4. pp. 561-577. (prior to chapter 2)

"Drumlins in County Down." Alan Hill. 1973. "The Distribution of Drumlins in County Down, Ireland." *Annals of the Association of American Geographers* 63:2. pp. 226-240. (prior to chapter 2)

"Patterns of American Agricultural Fairs." Fred Kniffen. 1949. "The American Agricultural Fair: The Pattern." *Annals of the Association of American Geographers* 39:4. pp. 264-282. (prior to chapter 3)

"Urban Heat Islands." Warner Terjung and Stella S-F. Louie. 1973. "Solar Radiation and Urban Heat Islands." *Annals of the Association of American Geographers* 63:2. pp. 181-207. (prior to chapter 4)

"Texas Long Lots." Terry Jordan. 1974. Antecedents of the Long-Lot in Texas. *Annals of the Association of American Geographers* 64:1. pp. 70-86. (prior to chapter 5)

"The Fisher-Victoria Tramway System in Louisiana." Michael Williams. 1989. *Americans and Their Forests*. Cambridge: Cambridge University Press. (prior to chapter 5)

"Percentage of the Sections Entered Under the Timber Culture Act" and "Percentage of Public Domain Entered Before 1872" (Nebraska, Kansas, Missouri, Illinois, and Minnesota). C. Barron McIntosh. 1975. "Use and Abuse of the Timber Culture Act." *Annals of the Association of American Geographers* 65:3. pp. 347-362. (prior to chapter 6)

"Movement of Summer Vacationists within Luce County Michigan, 1929." George F. Deasy. 1944. "The Tourist Industry in a "North Woods" County. *Economic Geography* 25:4. pp. 240-259. (prior to chapter 6)

"Meanders in Anatolian Rivers." Richard Russell. 1954. "Alluvial Morphology of Anatolian Rivers." *Annals of the Association of American Geographers* 44:4. pp. 363-391. (prior to chapter 7)

"Rockford, Illinois." John Alexander. 1952. "Rockford, Illinois: A Medium-Sized Manufacturing City." *Annals of the Association of American Geographers* 42:1. pp. 1-23. (prior to chapter 8)

"Cars of Malting Barley By County of Origin, 1939-40." John C. Weaver. 1944. "United States Malting Barley Production." *Annals of the Association of American Geographers* 34:2. pp. 97-131. (prior to chapter 8)

"Chorography of Southern New England - selected areas." Preston James. 1929. "The Blackstone Valley." *Annals of the Association of American Geographers* 19:2. pp. 67-92. (prior to chapter 9)

"Plan of Typical House - Nangodi, Ghana." John Hunter. 1967. "The Social Roots of Dispersed Settlement in Northern Ghana." *Annals of the Association of American Geographers* 57:2 pp. 338-349. (prior to chapter 10)

"Generalized Winter Storm Patterns" & "Meandering River." Robert De C. Ward. 1914. "The Weather Element in American Climates." *Annals of the Association of American Geographers* 4. pp. 3-54. (prior to chapter 10)

"Machine Space of Detroit, Michigan." Ronald Horvath. 1974. "Machine Space." *Geographical Review* 64:2. pp. 167-188. (prior to chapter 11)

"The Location of Agricultural Production after Von Thunen." Andreas Grotewold. 1959. "Von Thunen in Retrospect." *Economic Geography* 35:4. pp. 346-355. (prior to chapter 11)

"Unincorporated Hamlets in Wisconsin." Glenn T. Trewartha. 1943. "The Unincorporated Hamlet." *Annals of the Association of American Geographers* 33:1. pp. 32-81. (prior to chapter 12)

The quotes at the end of each chapter come from many sources, including P. & J. Muehrcke, "Maps in Literature" (*Geographical Review,* 64:3, 1974) and G. Brannon & L. Harding, *Carto-Quotes* (Upney Editions, 1996).

# Finally...

*Denis Wood:* I would like to thank Christine Baukus and Irv Coats for their continued support.

*John Krygier:* The origins of this book are in my experiences in the Department of Geography at the University of Wisconsin-Madison. Onno Brouwer gave me a job in the "Cart Lab," and there Onno, David Woodward, and David DiBiase all had a substantial influence on my development as a map maker. David Woodward, in particular, got me interested in map design and the history of cartography. He was a unique man and will be greatly missed.

Many thanks to several reviewers, and in particular Mark Monmonier, who made substantial comments on an earlier draft of the book. Reviewer insights are an important part of *Making Maps*. Ohio Wesleyan University and, in particular, Dean of Academic Affairs Richard Fusch have supported this project in many ways over the past few years. The *Globehead!* spirit pervades this book and thanks to those Globeheads who made my time at Penn State interesting. Denis Wood pulled the book out of the doldrums, and has been an inspirational coauthor. Finally, thanks to Patti and John Riley, who have put up with this book for a *long* time, and Annabelle who was delivered about the same time as the book.

Readers should inform John Krygier of errors, suggestions, and other comments related to this book: http://makingmaps.owu.edu/

# About the authors

John Krygier is a geographer with degrees from the University of Wisconsin–Madison and The Pennsylvania State University. He has extensive experience with map design and production, and has taught mapping and GIS at Penn State, the University of Oregon, The Ohio State University, and the University at Buffalo, The State University of New York. He is past president of the North American Cartographic Information Society, and will be the editor of the peer-reviewed journal *Cartographic Perspectives*. He teaches mapping, GIS, and geography at Ohio Wesleyan University, and lives in Columbus, Ohio.

Denis Wood curated the Smithsonian's award-winning Power of Maps exhibition and wrote the bestselling *The Power of Maps* (Guilford). More recently he has published *Seeing Through Maps* (ODT) and *Five Billion Years of Global Change: A History of the Land* (Guilford). He is an independent scholar living in Raleigh, North Carolina.